T0244415

'Mind-blowing. Everyone who cares about nature should read this book.'

James Rebanks, author of *The Shepherd's Life* and *English Pastoral*

'A very informed, impressive book. Essential for understanding the horrifying impact of roads and motor vehicles on nature.'

Derek Gow, author of *Bringing Back the Beaver*

'Remarkable! An immensely readable eye-opener. How could we have been so unaware of something so obvious and so damaging to wildlife?'

Tim Birkhead, author of *Birds and Us*

'*Traffication* is a book to slow down for: provocative, eye-opening and painstakingly researched. It's going to make me rethink the ways we impact our planet through one of the most simple of acts.'

Stephen Rutt, author of *The Eternal Season* and *The Seafarers*

'Paul Donald's *Traffication* is undoubtedly one of *the* environmental books of 2023. With perfect timing and tone, the author takes us through several essential learning curves and shows us how the car crisis, which most conservationists have long missed, is overwhelming large parts of nature. I could not recommend it more highly.'

Mark Cocker, author of *One Midsummer's Day*

'Every so often, a book comes along that has a profound impact on how we think and "do" transport. *Traffication* is one of those books, showing how the narrow focus on making car travel easier and faster is fundamentally harming the systems that wildlife depends on and restricting nature into tighter and tighter pockets. It's a really readable, clear and compelling case to put the countryside more at the heart of how we manage our transport system.'

Richard Hebditch, UK Director, Transport and Environment

'A brilliant and comprehensive expose of what roads are doing to our wildlife: meticulous, persuasive, challenging and brilliantly researched.'

Ben Macdonald, author of *Rebirding* and *Cornerstones*

'This book gives a well-researched and engagingly written account of what is arguably one of the major conservation issues of our time. In drawing attention to the greatly underestimated problems posed to wildlife and the wider environment by our ever-increasing road networks, traffic volumes and speeds, Paul Donald provides an important wake-up call, and importantly, discusses mitigating measures.'

Professor Ian Newton FRS, ornithologist and conservationist

'As the realisation of our treatment of the earth grows, a reassessment is underway, and *Traffication* adds a new and vital dimension. The benefits and the conveniences of the car are weighed against the devastating toll on wildlife and our own health and, increasingly, it doesn't add up – but is it possible to see a different future? This book says it is. A masterful analysis of a hugely important elephant-in-the-room topic, humanity's addiction to the car.'

Mary Colwell, author of *Curlew Moon* and *Beak, Tooth and Claw*

'A meticulously researched exposé of how we've been asleep at the wheel for years. This is a thought-provoking, and brave, examination of the damage we've caused that will hopefully jolt us from complacency and help us to modify our road-building and driving behaviour for the benefit of wildlife and human health. *Traffication* is the conservation conundrum we need to address with urgency.'

Dr Ruth Tingay, conservationist and co-director of Wild Justice

'We normally think of road transport as an urban problem but the creeping harm from traffic is suffocating our rural environment like an invasive species. This carefully researched book completely reframes the way that we should view traffic and highlights a blind spot for many conservation organisations.'

Dr Gary Fuller, author of *The Invisible Killer: The Rising Global Threat of Air Pollution*

'We know that traffic kills people through injuries, air pollution and inactivity, but Paul Donald shows with convincing science in his very readable book how, almost unnoticed, traffic has been destroying wildlife and the countryside. He shows too how we can take action that should not be painful.'

Richard Smith, chair of the UK Health Alliance on Climate Change

'*Traffication* tells the story of how quickly the car transformed our world and how, equally quickly, scientists highlighted the downsides. But despite several decades of growing evidence, the impact of traffic on the environment remains focused upon congestion, climate change and air pollution, while ignoring the more rural issues that impact directly on nature. The author offers beautiful, heartfelt writing and some hopeful concluding chapters.'

Baroness Jenny Jones, UK Green Party

TRAFFICATION

TRAFFICATION

How Cars Destroy Nature and What We Can Do About It

PAUL F. DONALD

PELAGIC PUBLISHING

Published by Pelagic Publishing
20–22 Wenlock Road
London N1 7GU, UK

www.pelagicpublishing.com

A CIP record for this book is available from the British Library

ISBN 978-1-78427-444-3 Hardback
ISBN 978-1-78427-445 0 ePub
ISBN 978-1-78427-446-7 ePDF

https://doi.org/10.53061/PXIN6821

Typeset in Chennai, India by S4Carlisle Publishing Services

Cover illustration by Jo Walker

Printed in the Czech Republic by Finidr

MIX
Paper | Supporting
responsible forestry
FSC
www.fsc.org FSC® C014138

Contents

Mirror, Signal, Manoeuvre

I grew up in a carless household. My parents have never driven, largely on aesthetic and environmental grounds. They have always considered the car to be a noisy, intrusive and largely unnecessary machine. They have proved the last point with a lifetime of walking, cycling and developing advanced skills in the interpretation of complex rural bus timetables. Their lives have not suffered at all as a result of being carless. If anything the opposite is true; the fact that they remain fit and active well into their 80s may be one of the rewards of their lifelong pedestrianism.

I do not have the moral fibre of my parents and learned to drive almost as soon as I was old enough, largely so that I could go birdwatching in ever more remote locations. But their dislike of the *infernal* combustion engine has engendered in me a lifelong sense of guilt for being a driver. So when in the 1990s a team of Dutch researchers published a series of scientific articles that demonstrated more clearly than anything before the profoundly damaging impact that road traffic has on our wildlife, I took note. Surely, I thought, this is going to be the start of something big in conservation. The UK has just as dense a network of roads as the Netherlands, and if the results of Rein Reijnen and Ruud Foppen's studies reflected conditions here, then our entire countryside must be at risk. Over the years I have kept a close eye on the scientific research linking road traffic to declines in wildlife populations, and have watched this new branch of science expand and mature since the turn of the millennium into a discipline that has gained sufficient momentum to warrant its own name: road ecology. In its short life, road ecology has built up a body of scientific evidence that permits only one conclusion: road traffic has wrought immense damage on the world's wild plants and animals, and it has done so in many different ways.

All the while I waited for the conservation world to sit up and take note of this mass of new scientific research, to open its eyes to the fact that road ecologists were finding more and more evidence to link our collapsing wildlife populations to a rising tide of road traffic. I saw issues such as agricultural intensification and climate change emerge as big new threats to wildlife, and contributed in a small way to research in these areas, and I watched from the inside as large conservation organisations rose to meet those challenges. But I am still waiting for the conservation movement, along with the general public, to wake up to the reality that road traffic poses threats to wildlife that are every bit as serious.

Then it occurred to me that perhaps part of the reason that roads are not generally seen as an existential problem for wildlife is that nobody has tried to pull all this new research together, to make a case that everybody can understand that here is something huge, something of global concern that somehow got overlooked. It seemed to be an inexplicable omission from the catalogue of popular science books. I know of at least ten popular books that have as their subject the impacts of agriculture on wildlife, and a similar number that discuss the impacts of climate change, yet as far as I am aware this is the first attempt at a book (in any country or language) that tries to synthesise in plain language all the many impacts of road traffic on the natural world.

My initial intention was to focus as much as possible on the situation in my home country, the UK, but for reasons that I cannot fathom the new science of road ecology has almost entirely bypassed us here. I have therefore had to discuss the many problems that our wildlife faces from road traffic using examples from countries where this new branch of science has really taken off over the last two decades. The USA, Canada, the Netherlands, Spain, Portugal, Sweden, Poland, Brazil, Australia and a number of other countries are all miles further down this avenue of research than we are. The UK enjoys a fine international reputation for the quality and quantity of its environmental science and I have no good explanation for why we lag so far behind in this particular field. On the plus side, this means that I have been able to draw on studies of a wonderful menagerie of non-British species: moose, grizzly bears, wolverines, rattlesnakes, giant anteaters, elephants, condors, tapirs, wombats, chimpanzees and tigers all have walk-on parts in this book about the UK's wildlife. Because what road ecology has taught us is that the principles are the same, whatever the country or the species.

This book is intended to be an examination of only the environmental costs and benefits (for there are a few of the latter) of roads and their traffic. These form just two of many columns in the car's overall balance sheet, and the book says very little about the many undoubted social and economic benefits of road traffic. Brian Ladd's book *Autophobia* (2008) reviews the history of passionate arguments that have been made for and against the car but sensibly shies away from taking sides. More recent histories, such as

Tom Standage's *A Brief History of Motion* (2021) and Bryan Appleyard's *The Car* (2022), offer largely positive accounts of our love affair with the motor car, although they are not entirely uncritical.

It would be a brave, perhaps foolish, person who tried to fill out the entire balance sheet and declare the car a net boon or public enemy, because it would be a near-impossible task. For a start, most of the benefits are unquantifiable – we can measure road accidents and air pollution with a good degree of precision, and we can count squashed badgers, hedgehogs and deer. But we cannot quantify benefits such as convenience or freedom of movement in the same empirical way. And even if we could somehow quantify the benefits as well as the costs, it would be impossible to equate the two – how many dead hedgehogs, or fatal traffic accidents, is one unit of convenience worth? And, just as important, to whom?

I have tried to find and interpret the evidence as impartially as my profession as a scientist dictates, but it has to be said from the outset that the prosecution has many more boxes of evidence on its desk than does the defence. That is just how it is; I don't suppose anyone opened this book expecting the car to emerge as our environmental saviour. *Traffication* is certainly not intended to be a polemic against the car (I am a driver myself), though such is the weight of environmental evidence against it that it might read like one in places. I have not attempted to balance accusations of environmental foul play with an appreciation of the many undoubted social and practical benefits of motorised transport. There have been many books written in praise of the car; this one simply attempts to balance the argument by examining some of the vehicle's less savoury aspects. A mature view of the world recognises that most things have both positive and negative facets, and that a detailed examination of one does not constitute a denial of the other.

While I use the word *car* in the subtitle of this book and frequently throughout the text, which may suggest a vendetta against that particular type of vehicle, in almost all cases I use it to refer to road traffic generally. My strong suspicion, backed up by some scientific evidence, is that heavy goods vehicles and motorbikes, being respectively larger and noisier, and faster and *much* noisier, than the average car, are individually far more damaging to the environment. But with the car making up over 80 per cent of all the vehicles on our roads, it makes a convenient scapegoat.

A note on units, definitions and data sources

The UK and the USA are the only countries of any size that record their national statistics in miles and use miles per hour (mph) as their unit of road speed. Most countries, and *all* scientific studies (including those written by authors from the UK and USA) use kilometres and kilometres per hour (kph) as their units of measurement. Rather than trying to standardise, I have used both imperial and metric measurements fairly interchangeably, depending on the context and the source of information. This is not ideal, but it seemed preferable to disguising the familiar 30 mph as a seemingly arbitrary 48.26 kph, or describing a carefully measured ten-kilometre transect from a scientific study as being 'about 6 miles long'. Sometimes one simply *sounds* better than the other, and here the imperial system generally wins out (in English, at least) through its long adoption into idiom.

One mile = 1.609 km, 1 km = 0.62 miles.

I have been similarly cavalier about road surfaces. Asphalt and tarmac are technically different substances (the first is bound by bitumen, the second by tar) and asphalt is used far more often in modern road building. But I have tended to use *tarmac* as a substitute for *road* when trying to avoid over-repetition of the latter; tarmac just sounds more 'roady' to me than asphalt, and it allows more alliterative opportunities.

Most of the road statistics quoted in the book come from the UK Government's Department for Transport annual statistics, which present information for Great Britain, rather than for the whole of the UK. Northern Ireland is therefore excluded from the data and so, with apologies, I refer to Britain rather than the UK when presenting or discussing these statistics. The government constantly revises the way it records road statistics and may apply changes retrospectively, meaning that the most up-to-date data may not always match up perfectly with the figures presented in this book.

Acknowledgements

I have been hugely lucky to have had a small focus group for this book who were kind enough to test-drive all the chapters and helped to improve their roadworthiness, often several times. They were my parents Trevor and Diana Donald, my friends Ken Allum and Jim Summers, and my wife Fiona Roberts. I am hugely grateful to them all. In the early stages of planning this book I received invaluable advice from Mark Cocker, who helped me get the engine started and pointed me in the right direction. For other comments and suggestions along the road I am very grateful to Dr Mark Avery, Professor Jeremy Wilson, Professor Nigel Collar, Alex Berryman, Dr Sam Gandy, Nathan and Ruby Rogers, Dr Sophia Cooke, Professor Andrew Balmford, Professor Rhys Green, Dr Ali Johnston, Dr Stuart Newson,

Professor Trevor Cox, Professor Clara Grilo, Professor Debbie Pain, Professor Kevin Gaston and Professor Tim Birkhead. Nigel Massen, Sarah Stott and David Hawkins at Pelagic have been a pleasure to work with and provided invaluable advice and support when most needed. Simon Fletcher did a great job of copyediting the final text.

Most of all I thank, again, my wife Fi for her belief in this project from the start, her unstinting support and her wonderful roadside recovery skills; whenever I broke down, she always managed to get me started again.

CHAPTER 1

The King of the Road

We need to go back a century and cross an ocean to reach the natural starting point for a book about the impacts of road traffic on Britain's natural environment. This brings us to Iowa City, USA, on the 13th day of the unusually wet June of 1924. That morning a zoology professor called Dayton Stoner and his wife Lillian, an ornithologist, packed some bags into their car, coaxed the engine into life and hit the road. Their destination was the Iowa Lakeside Laboratory on the shores of West Okoboji Lake, a drive of some 300 miles. Here the couple would spend a month teaching and studying before returning home.

The roads the Stoners travelled that day were surfaced with gravel or dirt and the couple seldom exceeded 25 mph, so each leg of their journey took them two days. Unlike Britain's meandering country roads – made, according to G.K. Chesterton's famous poem, by generations of rolling drunkards – Iowa's chessboard grid of rural highways is sober-straight, and the Stoners would have driven into a succession of long, linear vistas with only the occasional right-angle turn to trouble them. Unsurfaced country roads were usually shrouded in summer in a dense pall of dust, hurled back into the air by each passing vehicle, but the heavy rainfall of June 1924 had clarified the atmosphere and gummed the dirt to the ground. Conditions were perfect for viewing the highway ahead.

The Stoners had not travelled far before they became struck by the large number of dead animals they were seeing on the road, all clearly the victims of recent collisions with other motor vehicles. What made this otherwise utterly unremarkable road trip the natural starting point for our own journey was that instead of simply driving on by, Dayton and Lillian decided to identify and count all the little corpses they passed, stopping where necessary to examine the more mangled remains. Whether they did this simply to relieve the boredom of a long journey or whether they had an intimation of scientific immortality will never be known, but the following year the Stoners published their results in the prestigious journal *Science* and thereby unwittingly pioneered a whole new field of environmental science, known today as road ecology.[1]

'The Toll of the Automobile' was the first article ever published on the impacts of the car on wildlife, and listed the 225 dead reptiles, birds and mammals, of almost 30 species, counted by the Stoners along their way (their total also included 26 chickens and three pet cats). The most common casualty by far was the stunningly attractive red-headed woodpecker, a bird that was clearly very much more abundant in 1924 than it is today. Another species of woodpecker, the northern flicker, was also a frequent victim. As well as feeding in trees, like most woodpeckers, red-heads and flickers also feed on the ground and swoop low across roads to catch insects in flight, increasing their risk of being struck by passing vehicles. The most frequent mammalian casualty recorded by the Stoners was the thirteen-lined ground squirrel, a beautiful little rodent with intricate markings running along its back that look, somewhat un-fortunately, like five parallel roads (complete with dashed white lines down the middle).

The Stoners realised that casualty rates were highest along roads with better surfaces, which 'permit of greater speed, together with more comfort to the speeder and correspondingly greater danger to human and other lives'. By 'speeder', they meant those driving at more than 35 mph. Cars zipping along at this speed, the Stoners thought, gave animals feeding on the road, or crossing over it, little chance of escape. 'Assuming that these conditions prevail over the thousands of miles of improved highways in this state and throughout the United States', their article concluded, 'the death toll of the motor car becomes still more appalling.'

The Stoners' great insight was that road traffic 'demands recognition as one of the important checks upon the natural increase of many forms of life'. Thus the very first article ever published on the subject recognised that road traffic can affect populations of wild animals to the same extent as natural regulators such as predation, disease or starvation. This century-old observation has fallen on several subsequent generations of largely deaf ears.

Unbeknown to the Stoners, or indeed to anyone else, the car's ecocidal potential had already been spotted by another observant American zoologist. Joseph Grinnell, who worked largely in California, kept detailed notes on roadkill (and indeed on pretty much everything else he saw) in his private field journals from as early as 1920, although he never published them. His journal entry of 5 May 1920 anticipated the Stoners' insight by four years and expressed his concerns in remarkably similar language, even down to identifying the same wildlife-critical speed:

> A notable thing as one autos over the state highway is the number of dead animals of various sorts on the road. Yesterday noted: Jack Rabbit (many); Cottontail (many) . . . Kangaroo Rat; Bushey Ground Squirrel; Skunk; domestic dogs and cats; Meadowlark (2 or more); Bullock Oriole; Mockingbird. Even with my 24-mile an hour Ford, there were some close calls as regards some birds and a Jack Rabbit. With big machines, traveling 35 to 55 miles per hour, and with their intense lights at night, the animals happening to be on the road at night are in serious danger. This is a relatively new source of fatality; and if one were to estimate the entire mileage of such roads in this state [California], then mortality must amount into the hundreds and perhaps thousands every 24 hours.

Dayton Stoner (1883–1944, sitting just behind the front row, third from right) and his wife Lillian (1885–1978, two behind him at the back), photographed in 1918 with other members of the University of Iowa Barbados-Antigua Expedition. Dayton provided the expedition's evening entertainment by playing his mandolin. He was also threatened with being stoned to death by the locals, not for his mandolin playing but because they thought for some reason that he was a German spy. Lillian later became state ornithologist for New York. (*Biodiversity Heritage Library*)

The Stoners' article was quickly followed by a flood of similar publications, as scientists fell over themselves in a stampede to rush their observations of this emerging peril into print. Counting little corpses while driving slowly along quiet country roads is a fairly easy way to collect data, especially if you have to make the journey anyway, and the editors and readers of scientific periodicals seem to have had an almost insatiable appetite for these catalogues of death. In his entertaining history of roadkill research, Gary Kroll has termed this phenomenon 'dead-list mania', a craze for publishing inventories of roadside carcasses. Hard on the heels of the Stoners' original article came, among others, 'Is the Automobile Exterminating the Woodpecker?' (1926), 'Automobile Toll on the Oregon Highways' (1926), 'Feathered Victims of the Automobile' (1927), 'Speeding Motor Cars Take Toll of Wild Life' (1929 – the Stoners again), 'An August Day's Toll of Birds' Lives on Primary Iowa Roads' (1933), 'The Automobile as a Destroyer of Wild Life' (1934), 'The Death-roll of Birds on our Roads' (1936, the first British contribution on the subject), 'The Toll of Animal Life Exacted by Modern Civilisation' (1937) and 'Feathers and Fur on the Turnpike' (1938). My favourite title, which leaves few doubts as to the author's opinion of the automobile, has to be 'And Now the Devil-Wagon' (1926).

The Stoners had woken the world to the fact that a significant new threat to wildlife had arrived, one with the potential to slaughter huge numbers of wild animals over vast areas. But where had it come from?

Dawn of the Century

In 1900, with the Victorian era and its eponymous queen entering their final months of life, an extravagantly moustachioed American composer and music publisher called E.T. Paull released a rousing piano march entitled *Dawn of the Century*. The music itself has not stood the test of time particularly well (in fact, it's awful), but the illustration on the front cover of the sheet music is a striking and colourful example of *fin-de-siècle* confidence that has been widely reproduced.

Progress is embodied in the form of a young woman (perhaps Columbia) in a risqué robe of billowing silk. She is standing on a winged wheel, the coming daybreak behind her all pinks and oranges. In her left hand she holds aloft a standard on which is written 'XX Century' (modernity still clearly preferring the use of Latin numerals). Her right hand hovers over a telegraph key that radiates lightning bolts to illustrate the speed and range of communication available to her. Rather inelegantly, she has an electric light bulb strapped to her forehead. Around her fly images of progress – an

Dawn of the Century (1900). Spot the car. *(Wikimedia Commons)*

electric tram, a telephone, a steam locomotive, a reaping machine, a camera, a sewing machine. And there, emerging from behind the hem of her gown, is a car. But while the tram and the locomotive race confidently forwards across a brightening orange sky with headlights blazing, the car, unlit and shrouded in dark clouds, is pictured back-on, as though embarrassed to be included in such progressive company. Compared with the brazen assurance of the other symbols of humanity's advancement, the car appears to have been added almost apologetically.

The unnamed artist would not have been the only one at the time to wonder whether this newfangled contraption was a fitting icon for the dawning century, for its contribution to the sum of human health, wealth and happiness had thus far been negligible (and on the first two scores at least, largely negative). In the year that *Dawn of the Century* was published, sales of new cars in the USA – most of them comical boneshakers powered by steam or heavy lead-acid batteries – numbered just 4,000, one for every 20,000 inhabitants. In Britain there were fewer than 800 horseless carriages on the roads and commercial manufacturing was in its infancy. Most of the world's countries had no cars at all.

France was by far the largest car manufacturer, producing over half the world's motor vehicles. Its domination of the early car industry has left a lasting legacy in the vocabulary of motoring. Gallic words such as chauffeur (from *chauffer*, someone who stokes a steam engine), chassis, garage, limousine, coupé and carburettor all point back to the original centre of mass car manufacture. It was also in the vehicle workshops of Paris that a new word was forged by welding a Latinate rear onto a Greek front end – automobile.

Practically everything that moved on the roads, in the towns and in the countryside, was pulled by horses, of which there were over 3 million in Britain in 1900 (a tenth of them working on the streets of London). Between 1870 and 1900, the number of working horses in European and American cities increased greatly to service the long-distance commerce offered by the railways: trains were useless if nobody could get themselves or their produce to the station. This reliance on horse-drawn travel persisted in rural areas well into the age of motoring. Laurie Lee, in *Cider with Rosie*, described how just after the First World War:

> The horse was king, and almost everything grew around
> him: fodder, smithies, stables, paddocks, distances and
> the rhythm of our days. His eight miles an hour was the
> limit of our movements, as it had been since the days
> of the Romans. That eight miles an hour was life and
> death, the size of our world, our prison.

If the horse was the king of the road at the end of the Victorian age, then the bicycle was its prince. Many roads, even those linking major cities, had fallen largely into disuse as the railways and canals dominated the carriage of people and goods. Commercial intercity stagecoach services had largely ceased by the 1860s, freeing up the roads for other users. These quiet, inviting highways, coupled with the invention in the 1880s of the pneumatic tyre and the safety bicycle (often just called the 'safety', to distinguish it from the 'ordinary', as the ludicrous penny-farthing was known), sparked a late Victorian vogue for cycling. The bicycle's popularity has endured periods of boom and bust since its invention, but the cycling mania of the 1890s was unlike anything seen before or since. Cycling appealed to all social classes as an inexpensive form of emancipation that allowed people to escape to the countryside without the rigid prescription of railway timetables or the cost of maintaining a horse and carriage. In 1897, there were over 2,000 cycling clubs in Britain, 300 of them in London alone.

British bicycle manufacturing boomed during the 1890s, employing up to 50,000 workers. Share prices of bicycle companies tripled in the space of a few months in 1896, and the number of bicycle companies expanded more than fivefold: between 1895 and 1897, nearly 700 cycle companies were floated on the stock exchange. It was, in economic terms, a bubble. The products of companies such as Raleigh and Singer were exported all over the world. Cycling in the USA enjoyed a similar boom in popularity; by 1896, Chicago's factories were turning out a quarter of a million bicycles each year and American manufacture would soon rival British output. Demand for bicycles and tricycles, it seemed, was almost insatiable.

As Carlton Reid has persuasively argued in his book *Roads Were Not Built For Cars* (2015), the cycling craze of the 1890s paved the way (literally and figuratively) for the later success of the motor car. First, it opened people's minds to the possibility that roads were not only routes of commerce, but could also be avenues of leisure and pleasure. It was the large and influential cycling lobby that pushed for the upgrading of road surfaces to accommodate their wheels, an improvement soon to be rudely appropriated by the car. Second, the engineering expertise that was built up through the mass production of cycles laid the mechanical foundations of the early car industry. Some of the earliest cars, such as Carl Benz's Patent-Motorwagen (1886, arguably the world's first production automobile) or the Wolseley Autocar Number One (1896), were little more than motorised tricycles. William Morris (later Lord Nuffield), who founded Morris Motors, began his career making bicycles, and the companies of Rover, Humber and Singer also started life as cycle manufacturers before venturing into the car market. Henry Ford's first motorised vehicle, the Quadricycle (1896), was built largely of bicycle parts, and he remained a keen cyclist throughout his life.

The similarity between the first luxury cars and horse-drawn carriages was superficial, for beneath the leather upholstery and gilded coachwork lay very different machines. 'If a paternity test were possible', Carlton Reid wryly observes, 'it could be shown that the first motor cars had much more cycle DNA in them than carriage DNA.' It was no coincidence that the main centres of bicycle production – Paris, Detroit and Coventry – would soon become the heart of motor manufacture.

The bicycle's popularity during the last few years of Victoria's reign is hard to overstate. An article published in the cycling paper *The Clarion* in 1897 claimed:

> The man of the day is the Cyclist. The press, the public,
> the pulpit, the faculty, all discuss him. They discuss
> his health, his feet, his shoes, his speed, his cap, his
> knickers, his handle-bars, his axle, his ball-bearings, his
> tyres, his rims, and everything that is his, down unto his
> shirt. He is the man of *Fin de Cycle* – I mean *Siècle*. He is
> the King of the Road.

This account is hardly gender neutral, but it was published in a newspaper that did much to promote cycling for all as part of its egalitarian agenda. *The Clarion* established a large number of cycling clubs in working-class areas, and unlike others it actively extended membership to women when it founded its first chapter in 1894; the 'Clarionettes' included among their members the suffragettes Christabel and Sylvia Pankhurst. The cycling mania of the 1890s contributed much to the empowerment and liberation of women; in 1898, Susan, Countess of Malmesbury, a well-known cycling writer, pronounced it 'one of the greatest blessings given to modern women'.

The rapidity with which the bicycle's popularity grew in the mid-1890s was summed up by another female cycling writer, Constance Everett-Green: 'It is not too much to say that in April of 1895 one was considered eccentric for riding a bicycle, whilst by the end of June eccentricity rested with those who did not ride.'

But just a few weeks after eccentricity decided to swap sides, an event took place that would mark the beginning of the end of the bicycle's chances of usurping the horse as king of the road. Because it was in July 1895 that the real '*Fin de Cycle*' first came chugging down a British lane, scattering chickens and children before it and leaving the past shattered in its wake.

'Our iron horse behaved splendidly'

The pretty village of Micheldever in England's southern county of Hampshire boasts a fine example of an early Victorian railway station, neatly faced in knapped flint and looking not much different now from when it first opened its ticket office in 1840. Unlike so many other small rural stations, it survived the savage cuts of the 1960s and remains in use. Passengers waiting on the platform today might notice a smart red plaque on the wall, unveiled by the National Transport Trust in 2021, which commemorates an event that took place here in July 1895. That month, the Hon. Evelyn Ellis imported from Paris a Panhard et Levassor motor carriage, built to his own specifications and powered by a twin-cylinder, four-horsepower Daimler engine. The firm of Panhard et Levassor was one of the biggest and most prestigious manufacturers of its day and is often credited with building cars that were to set the industry standard, having a front-mounted radiator and engine, rear-wheel drive and something not dissimilar to a modern transmission to link them.[2]

The most widely recounted version of the story tells that Ellis collected his new car in Paris and drove it to the Channel port of Le Havre, then crossed by boat to Southampton, from where the car was transferred by train to Micheldever station. Others have suggested that Ellis drove his car from Southampton to Micheldever. Either way, what is generally agreed is that on the morning of 5 July 1895, Ellis set out from Micheldever to drive to his home at Datchet in Berkshire. Whether it started in Southampton or in Micheldever, Ellis's journey was the first ever undertaken in Britain in a combustion-engine car. It marked the beginning of British motoring, and the beginning of the end of cycling's brief golden age.

Ellis's companion on that historic day was the engineer Frederick R. Simms, an influential figure in the story of British car manufacture. Simms was a close friend of Gottlieb Daimler and in 1890 had acquired from him the rights to manufacture and sell, 'in England and the colonies', Daimler's petrol engine. Among his many achievements, Simms founded the Automobile Club of Great Britain (later the Royal Automobile Club, or RAC) and the Society of Motor Manufacturers and Traders; he also invented the ignition magneto and the armoured car, and is credited with coining the words *petrol* and *motorcar*. As if these were not achievements enough, he also discovered, and perhaps even created (by diverting a watercourse), a spectacular waterfall in the Austrian Alps still known today as the Simms-wasserfall. Simms later sold his Daimler rights to the scandal-plagued Harry Lawson, who in 1896 acquired a large factory in Coventry that he christened Motor Mills, Britain's first large-scale car production plant.

Simms described his famous journey with Ellis in an article for the *Saturday Review* entitled 'A Trip in a Road Locomotive':

> During the previous night a long and much-wanted
> steady rainfall had laid the dust on the roads, and thus
> we had every prospect of an enjoyable journey. We set
> forth at exactly 9.26 a.m., and made good progress on
> the well-made old London coaching road … It was
> a very pleasing sensation to go along the delightful
> roads towards Virginia Water at speeds varying from
> three to twenty miles per hour. Our iron horse behaved
> splendidly. There we took our luncheon and fed our
> engine with a little oil … Going down the steep hill
> leading to Windsor, we passed through Datchet, and
> arrived right in front of the entrance hall of Mr Ellis's
> house at Datchet at 5.40, thus completing our most
> enjoyable journey of fifty-six miles, the first ever made
> by a petroleum motor carriage in this country, in
> 5 hours 32 minutes, exclusive of stoppages … In every
> place we passed through we were not unnaturally
> the objects of a great deal of curiosity. Whole villages
> turned out to behold, open mouthed, the new marvel
> of locomotion. The departure of coaches was delayed to
> enable their passengers to have a look at our horseless
> vehicle, while cyclists would stop to gaze enviously at us
> as we surmounted with ease some long and (to them)
> tiring hill … Mr Ellis's Daimler motor carriage … is a
> neat and compact four-wheeled dog-cart with accom-
> modation for four persons and two portmanteaux. The
> consumption of petroleum is little over a halfpenny per
> mile and there is no smoke, heat or smell, the carriage
> running smoothly and without any vibration.

Simms was at pains to stress to his largely equestrian readership that of the 133 carefully counted horses they encountered along the way, 'only two little ponies did not seem to appreciate the innovation'.

This historic journey was completed at an average speed of 10 mph. At such a breakneck pace the pair ran the risk of arrest, and indeed they even courted it, because Ellis had set out deliberately to flout a law that imposed a restrictive speed limit and required a guard (with or without a red flag) to walk in front of any self-propelled vehicle on a public road. The 1865 Locomotive Act, more commonly known as the Red Flag Act, set a maximum speed limit of 4 mph in the country, and just 2 mph in towns. In the event, Ellis and his companion managed to evade (or perhaps outrun) the forces of the law.[3]

A few months after Ellis's illegal sprint, Britain's pioneering motorists gathered at the Tunbridge Wells Agricultural Show Ground for a two-hour event billed as The Horseless Carriage Show, perhaps the first motor show ever held in any country. The exhibits comprised just five vehicles: two cars (one of them Ellis's), a motor tricycle, a steam tractor and a motorised fire engine. This might sound like a fairly underwhelming display for the 5,000 or more spectators who paid a shilling to enter, but it is hard for us now to imagine the profound impact that even these few vehicles would have had at the time. It turned out to be a hugely influential event; photographs of top-hatted Victorian gentlemen sitting stiffly in their imported French motor carriages, surrounded by crowds of enthralled onlookers, did much to stoke discussion about the possibility of horseless road transport in Britain. The newly founded *The Autocar* magazine enthused that 'this exhibitive trial will rank for this class of vehicle very much on an equality with that memorable trial of locomotives, in which the famous old Rocket so completely defeated the engines opposed to it at Rainhill just 66 years ago'. The vehicles laboured over the soft turf of the Show Ground and struggled to give their best performance, so the event's organiser, Sir David Salomons, 'steeled himself to dare the majesty of the law … for the carriages left the ring and came out upon the excellently laid highway which stretches between the showground and the town'. There the vehicles really came into their own, accelerating to speeds of 10 or 15 mph and thrilling the assembled crowd. Salomons, like Ellis a few months before, wanted to demonstrate the safety and roadworthiness of these new vehicles to the adoring crowd, and was prepared to break the law to do so. The fact that he happened to be the Mayor of Tunbridge Wells no doubt gave him confidence that he would not be arrested that day.

The following year, the government bowed to growing pressure and the Red Flag Act was replaced by the Locomotives on Highways Act, which raised the limit to 14 mph, the speed of a cantering horse. This freed up the highways for use by motorists and seeded a boom in car ownership that has never looked back.

A few weeks after the passing of the 1896 Act, three petrol-powered cars – of perhaps just a dozen such vehicles in the whole country – were brought into the centre of London, giving most people in the capital their first glimpse of this new invention. The Anglo-French Motor Carriage Company offered short, 4-mph demonstration rides around the grounds of the Crystal Palace to anyone willing to pay a shilling for the pleasure. On 17 August 1896, one of these cars struck and killed Mrs Bridget Driscoll, making her Britain's first motor-car accident fatality. The coroner at her inquest expressed his fervent hope that such a tragedy would never occur again.

In celebration of the repeal of the Red Flag Act, and despite the death of Mrs Driscoll, the recently formed Motor Car Club of Britain, with the louche self-publicist Harry Lawson at its helm, organised the November 1896 Emancipation Run from London to Brighton. Contemporary accounts suggest this was a chaotic affair, fuelled as much by alcohol as by petroleum spirit; one eyewitness watching the start, in thick smog, from the Hotel Metropole (where many of the contestants had enjoyed a largely liquid breakfast), wondered whether any of the drivers would come back alive.

The French, unsurprisingly, carried the day. First to arrive in Brighton, in a time of three and three-quarter hours, was Léon Bollée, followed an

The Hon. Evelyn Ellis, Britain's first petrol-powered motorist, in his French Panhard et Levassor. Like most of the earliest cars, it was steered with a tiller rather than a wheel. One of the modifications Ellis requested was that the controls of his car be shifted to the left; in the days before overtaking, it was usual for the driver to sit on the side of the vehicle closest to the kerb. Remarkably, the car survives to this day and can be seen in London's Science Museum. *(Heritage Image Partnership Ltd/Alamy)*

hour later by his brother Camille. A Panhard Wagonette came home in third place after five hours. After another long gap there was a thrilling race for fourth spot with no fewer than seven vehicles, including an electric bath chair, trundling over the finish line within the space of just a few minutes (it is likely that they had all hitched a ride on the same train from London). Evelyn Ellis suffered mechanical problems and did not reach Brighton until 3 a.m. the following morning, long after the earlier finishers had staggered to bed to sleep off their celebratory dinner. Many of the eager competitors who set off from London that morning never set eyes on Brighton at all, but the staging inns along the way did brisk business. It was a shambles: another eyewitness considered that 'there can have been very few converts to the cause of automobilism, and the numbers of vehicles which either broke down or were only able to limp into Brighton a long time after the arrival of the leaders served simply to increase the scepticism of those who had no faith in the future of the motor-car'. Yet the Emancipation Run somehow managed to inveigle its way into the canon of great motoring events; the 'Red Flag Run', as it became known, is celebrated to this day (in much more sober form) as the annual London to Brighton Veteran Car Run.

'A gay and meretricious swindle from first to last'

At the turn of the twentieth century, cars were used primarily for pleasure. They were a status symbol rather than a necessity, expensive toys far beyond the pockets of most people. Motorists were usually former cyclists with enough money to take the pleasures of the open road to new heights. At this early stage of motoring the car industry was highly fragmented, with many small independent manufacturers each producing just a handful of vehicles each year, and it had yet to reach consensus even on the best source of power for its machines. More than three-quarters of the four thousand or so vehicles sold in the USA in 1900 were powered by battery or steam, each of which came with its own set of problems.[4]

Electric vehicles were in commercial production from around the same time as the earliest petrol cars, the first of them perhaps the German Flocken Elektrowagen (1888). They were easier to drive than petrol or steam cars and, having fewer moving parts, were much less likely to break down. Battery-powered cars were also quiet, easy to start and produced no noxious smoke, and the high torque of electric motors gave them good acceleration and, for the day, high speeds. However, they were very expensive (a 1910 Waverley Electric Coupé, a tiny battery-powered sentry box on wheels, cost five times the average yearly salary of an American worker) and they had a limited range, perhaps only 20 miles or so. Opportunities for recharging batteries were equally limited, because in 1900 the domestic electricity supply was almost as rare as the car. Even if a supply could be found, the electric motorist then faced a lengthy wait while the heavy lead-acid batteries slowly regained their charge. Some pioneers of electric vehicles, such as the operators of the short-lived Electrobat taxi company, whose vehicles plied the streets of Manhattan in 1897, or the equally ephemeral London Electrical Cab Company (1897–9), hit upon the idea of using replaceable batteries, allowing the driver to whip out a flat battery and fit a freshly charged one in seconds. But for reasons that included physics, finance and fraud the vision never took off, and an aggressively pro-petrol journal called *The Horseless Age* was able to scoff at the concept's collapse.

The market for electric cars continued for a brief, sexist, period when they were marketed as vehicles for women – simple to drive, free of the need for crank-starting, reliable and suitable for short journeys to visit friends (but no further). To complete the patronising, some of these cars were designed with the driver facing her passengers, the better to chat with them (though not to see the road ahead), and were steered with a tiller, deemed to be easier for the weaker sex to turn even though it was less safe than a wheel. Henry Ford chose to buy his wife Clara an electric car rather than give her any of the 15 million Model Ts his factories produced. The fact that the first long-distance journey in any car had been undertaken by a woman had quickly been forgotten, yet the 120-mile round trip from Mannheim to Pforzheim by the intrepid Bertha Benz and her two young sons in 1888 remains one of motoring's truly great feats.

Ford's own efforts to produce an electric vehicle for the masses came to nothing, despite the involvement of his brilliant friend Thomas Edison. The first age of the electric car ended in failure and prejudice, and by the 1920s few were still in production. Their demise would haunt the industry like the ghost of a missed opportunity through many a city smog and oil crisis until the invention of the lithium-ion battery opened up a new opportunity for electric travel.

Steam was similarly impractical, and considerably more dangerous. Steam-driven cars lost up to a gallon of water per mile, requiring frequent refills, and building up steam pressure at the start of the journey could be a lengthy process. More than a few drivers of steam-powered cars were killed or mutilated when their boilers exploded. Rudyard Kipling, a grimly determined early devotee of motoring, described journeys in his American steam-driven Locomobile ('a gay and meretricious swindle from first to last') as a series of 'agonies, shames, delays, rages, chills, parboilings, road-walkings, water-drawings, burns and starvations'. How many of these indignities Kipling endured himself is unclear because like most motorists of the time he employed a chauffeur, who presumably bore the brunt of all these misadventures. Kipling tolerated his Locomobile for less than a year before purchasing a petrol-driven Lanchester, one of the few cars being produced commercially in Britain at the time, although it appears to have pleased him little better.

Like the first generation of electric cars, steam-powered cars waned in popularity through the first two decades of the twentieth century, and their commercial manufacture had all but ceased by the mid-1920s, although steam-powered goods vehicles, capable of pulling greater weights than any petrol engine of the time could manage, survived a couple of decades longer.

The main problem with the earliest internal combustion engines was to get them started in the first place. Cranking the engine manually was a strenuous and sometimes dangerous exercise; engines often backfired during cranking, throwing the person operating the handle to the floor or even breaking their arm. Furthermore, the refining of petroleum for vehicles had barely begun in 1900 and there were no petrol stations – fuel suitable for cars was sold largely by hardware shops as a cleaning agent. The first filling station in Britain did not open until 1919, at Aldermaston in Berkshire, its single pump, operated by Automobile Association (AA) patrolmen in full livery, dispensing a coal-tar derivative called benzole. The station's aim was to promote British-made benzole, a fuel that had previously been imported from Russia. The Russian Revolution of 1917 had interrupted supplies and tainted imported benzole with the stain of Bolshevism, a movement that was politically intolerable to the wealthy elite of British motorists. In addition to its pump, the Aldermaston filling station boasted a compressor for inflating tyres, a fire extinguisher and a toilet (described as being 'of bucket type, and seldom used').

Despite these teething problems, combustion would win the race. The invention of the starter motor in 1903 and the success of the Oldsmobile Curved Dash, the first petrol car to be produced in substantial numbers, ensured that the internal combustion engine would beat the battery and the boiler to dominate the market for over a century. Only recently has that domination started to be challenged.

'Poop-poop!'

The first cars suffered frequent breakdowns; parts were badly made and even core components such as axles commonly snapped. Rubber manufacture was an infant industry, and punctures were so frequent that motorists were forced to carry several spares. The job of changing a damaged tyre was difficult, dirty and exasperating. The poor state of many roads exacerbated these problems, and dust was a serious problem in dry weather: most vehicles of the era were not covered and many lacked even a windscreen. This led to the development of a form of dress that came to epitomise, even to caricature, early motorists: a sandy-coloured 'duster' overcoat and goggles, a driving cap for men, and for women a broad-brimmed hat tied in place under the chin with long gauze veils like a beekeeper's hood. Wet weather required the donning of a rubber cape.

The mechanical trials and tribulations of motoring pioneers were accompanied by no little animosity from those who objected to these new machines. Early motorists spent much of their time on the road (in both senses) repairing their vehicles and were a favourite object for jokes and jibes. Those whose livings depended on the horse were particularly sensitive to these unwelcome newcomers. Kipling, and presumably also his long-suffering chauffeur, endured numerous humiliations at the hands of jeering horse-cab drivers during their frequent breakdowns. Antipathy towards cars was widespread and often turned to violence. An irate horseman complained in a letter to *The Times* in 1901 about being overtaken by a car whose driver 'did not even slacken his pace; and my only consolation was that as the car swept by I brought the lash of my whip with all the force at my disposal across the shoulders of the driver and the man sitting by his side'. (One wonders how fast the car can have been travelling for the mounted esquire to be able to land such a blow.)

Another recurrent theme in Kipling's writings, and in early motoring literature generally, was the animosity shown by motorists towards the police (and vice versa). Speed traps were common, with concealed policemen using stopwatches to time vehicles along pre-measured distances. One of the most famous early motorists to fall foul of the law was an amphibian. Mr Toad, a character in Kenneth Grahame's evergreen novel *The Wind in the Willows* (1908), was a clever caricature of the wealthy, self-indulgent pleasure-seekers who epitomised motoring in its early years. It is clear that Grahame was no fan of the car because the book is, at heart, a thinly disguised attack on motoring, contrasting the antisocial petrol engine and the brashness of motorists with the virtues of more sedate forms of travel, such as rowing boats and horse-drawn caravans. In the novel, Toad is so instantly obsessed with the car that after his first disastrous encounter with one all he can do is sit in the middle of the road repeating, mesmerised, the sound of the fast-disappearing vehicle's horn: 'Poop-poop!' Toad's friends become so concerned by his sociopathic motoring that they place him under house arrest to protect him and others. But he escapes, steals a car and is then caught, tried and sentenced to twenty years in prison. He escapes again, this time by hoodwinking the gaoler's naïve daughter, but at the end of the story, after many further trials and tribulations, he repents his motoring madness and compensates those he has wronged.

Repentance was not the usual reaction of non-amphibian motorists who got into trouble with the police. Their wealth and high status meant that drivers tended to regard the enforcement of vulgar traffic laws as an affront, adding social insult to their many mechanical injuries. In 1903, Alfred Dunhill, who became wealthy pandering to the whims and fashions of well-heeled motorists, offered for sale his 'bobby finders', binoculars disguised as driving goggles that were guaranteed to 'spot a policeman at half a mile, even if disguised as a respectable man'.

The AA was established in 1905 largely to help motorists avoid speed traps, but its bicycle-mounted patrols had to take account of the fact that warning speeding drivers of these traps was itself illegal. A tacit agreement therefore emerged: if an AA patrolman failed to salute the driver of a car displaying a membership badge, it was a coded signal of trouble on the road ahead. The *AA Handbook* somewhat disingenuously informed its members that 'it cannot be too strongly emphasised that when a patrol fails to salute, the member should stop and ask the reason why, as it is certain that the patrol has something of importance to communicate'. The requirement for AA patrolmen to salute the organisation's members on the road continued until 1962 when it was stopped on the grounds of safety, the majority of patrols by then being mounted on motorbikes.

The first motorists faced condemnation and even violence from both sides of the social divide. The rural poor of Europe and North America were horrified by the slaughter made by passing cars of their pets, their livestock and, not infrequently, their children. Added to this was the frightening noise made by these vehicles and the suffocating dust they churned up as they sped through villages, which hung in the air long after the cough of the vehicle's engine had faded into the distance. Angry locals responded to this alien intrusion by throwing stones, digging trenches across roads, taking potshots at passing vehicles and even stretching rope or wire taut across the highway (several gory decapitations were recorded). Writing of a car journey through the Netherlands in 1905, a German motorist recorded that 'most of the rural population hates motorists fanatically', and that she 'encountered older men, their faces contorted in anger, who, without any provocation, threw fist-sized stones at us'. At a convention held in Montana in 1909, the Farmers' Anti-Automobile League called on its members to 'give up Sunday to chasing automobiles, shooting and shouting at them'. For their part, wealthy motorists blamed the peasantry for not understanding the rules of the road, and tended to regard the collision of fender with flesh as being the fault of the victim for not getting out of the way in time. No wonder Dorothy Levitt, author of *The Woman and the Car* (1909) and an influential early British motorist in her own right (she invented the rear-view mirror), recommended that female drivers always carry a small pistol with them on their travels.

At the other end of the social scale were members of the conservative establishment and the reactionary intellectual elite who disliked motoring largely because it upset the old equestrian order. They saw road safety as a convenient stick with which to beat the car. In Britain, the conservative, horse-loving Highway Protection League (initially the Pedestrians' Protection League) was established in 1902 by the Tory peer Lord Leigh for 'the protection of the public from excessive motoring speed and the introduction of legislation to remove the terrorism of the flying motor'. They used dramatic images of small children frozen in fear on the road in front of speeding cars, their goggled drivers hunched maniacally over the wheel, to drum up support for the imposition of strict speed limits. Following Lord Leigh's death shortly afterwards, the presidency of the short-lived League passed to Richard Verney, 19th Baron Willoughby de Broke, described rather uncharitably (though not entirely inaccurately) as a man 'whose face bore a pleasing resemblance to the horse', and who was 'not more than two hundred years behind his time'. In 1909, the League's Chairman wrote a letter to the editor of the *Manchester Guardian* protesting 'against a proposal to sacrifice the comfort and safety of a vast majority to the caprice of a very small though wealthy and powerful minority'. But the League slowly dissolved as more and more of its members succumbed to the allure of motoring.

Sporadic protests persisted long after the car had won the day. The German sociologist and economist Werner Sombart complained bitterly of a world in which 'one person was permitted to spoil thousands of walkers' enjoyment of nature'. In 1927, the eccentric English philosopher and early celebrity broadcaster C.E.M. Joad continued to rail against the car, maintaining that 'motoring is one of the most contemptible soul-destroying and devitalizing pursuits that the ill-fortune of misguided humanity has ever imposed upon its credulity'.[5]

The early pioneers of motoring, then, suffered numerous mechanical tribulations and faced hostility from many quarters. Their vehicles were expensive, slow, unpopular, dangerous and capricious, and ran on roads designed more for hoofed than for wheeled transport. Very few people owned one, or could ever hope to; indeed many people, even in the most motorised countries, had still never even seen one. The car appeared to pose so little threat to more established forms of transport that during the annual industry gathering of horse-carriage manufacturers in 1900 it was not mentioned once. It is hardly surprising that the artist who created the front cover of *Dawn of the Century* appeared so uncertain about depicting the car among other more established symbols of progress.

There was nothing to suggest, as the twentieth century opened for business, that the motor car would ever be more than the noisy plaything of a wealthy few. Woodpeckers fed undisturbed by the roadside; rabbits and hares lolloped safely across. But the peace was about to be shattered, and the countryside changed forever.

CHAPTER 2

Traffication

In a remarkably short period of time, between the bravura artwork of
Dawn of the Century and the Stoners' corpse-strewn journey across rural
Iowa a quarter of a century later, the car became king. Someone born in
Western Europe or North America in 1900, when motorists comprised a
few wealthy hobbyists in preposterous steaming contraptions, would still be
young in an age of conveyor-belt production, multi-storey car parks, long-
distance commuting and lengthy traffic jams. In a single human generation,
the hobby of a privileged elite became the birthright of the masses, and it
utterly transformed how and where people lived.

In a year that saw the launch of the ill-fated *Titanic*, the storming of
Parliament by suffragettes and the crowning of King George V as Emperor
of India, one anonymous Londoner was more concerned by the growth of
motoring:

> It seems but yesterday that public opinion all over the
> country was gradually waking up, holding up its hands
> in wonder, and saying, 'The motor-car has come to stay.'
> And all the while the motor-car has been not staying,
> but pursuing its inevitable way, imposing itself upon the
> world in ways both fortunate and unfortunate. It began
> by being a scientific experiment, went on to become
> the instrument of the adventurous, then became the toy
> of the rich, then the ambition of the poor, and finally
> the servant of everyone. Ten years ago it was a fantastic
> luxury, and today it is a dire necessity. From being the
> plaything of society it has come to dominate society.

It seems extraordinary that these words could have been written as early
as 1911, just 16 years after the frock-coated avant-garde of motoring had
gathered five primitive vehicles together in a muddy field in Tunbridge
Wells. The Horseless Carriage Show of 1895 had featured only imported

French motor cars because there was no British manufacturing. Just ten years later, the 1905 Motor Show at London's Olympia exhibition centre could boast a range of British-made vehicles that claimed to match any foreign rival. From fewer than 800 cars on Britain's roads in 1900, many of them still powered by steam or battery, the number had risen to 23,000 by 1904 and to more than 60,000 by 1910. The number would exceed a million by the early 1930s and the second million would be added before that decade was out. The Second World War hit civilian motoring hard, and the number of cars on our roads in 1950 was no higher than it was in 1940. But since then the number of vehicles has followed only one course: 2 million cars in 1950 would double to 4 million by 1958, double again by the mid-1960s, double again by the early 1980s – and then double once more to the more than 30 million cars we have on our roads today.

A book entitled *Man and the Motor Car*, published in 1936 by the Board of Education of the City of Detroit,[6] summarised (in rather more positive terms than the grumpy Londoner of 1911) what the car had become:

> The man of today probably travels at least a hundred
> times as far during his lifetime as the man of fifty years
> ago … Automobiles are such a common part of our
> social life that we often disregard the great variety of
> services they perform. We know that they are used
> for everyday business and errands, for short pleasure
> trips and long vacation tours. We know that they carry
> doctors, lawyers, business men and workers to and from
> their jobs, that they move farm produce and general
> merchandise, from butter and eggs to steel and lumber.
> We know that under their influence towns have spread
> out and sprung up, that a man can work in one city, live
> in another and have his close friends in a third. More
> and more people quit city life, and choose life in the
> suburbs and the country. Workers and their families
> move from place to place and we all enjoy a new
> freedom. We are a nation on wheels.

The author of this book, which was aimed primarily at educating young drivers in road safety and etiquette, was not entirely uncritical of the car, noting for example that it 'has given criminals a new tool for their lawless work, one which is of more advantage to them than the machine gun'. More worrying still, 'the destruction of life, limb and property caused by our automobiles every year is appalling'. But by the time this book was written, all serious opposition to the car had been quashed. Even the Great Depression of the early 1930s did little to cool this new ardour; to the dismay of the mortgage companies, families falling on hard times often preferred to give up their homes before surrendering their cars, which at least allowed them to keep moving on in their desperate search for work.

'I will build a motor car for the great multitude'

This extraordinary revolution in how we travel, and indeed in how we live, is often attributed to one man: Henry Ford. This is rather an oversimplification – the first car produced using something akin to production line technology was the Oldsmobile Curved Dash (1901–4) – but there can be no doubt that Ford's approach to car manufacture was instrumental in bringing about one of the biggest social revolutions in our modern history. Ford recognised that the key to success, and therefore profitability, was not to pander to the wealthy few, as previous manufacturers had done, but to cater for the masses. Throughout his life he eschewed the luxury market, seeking instead to maximise his output of cheaper vehicles and rely on economies of scale to bring prices down while keeping profits high.

Ford outlined his credo in messianic style:

> I will build a motor car for the great multitude. It will
> be large enough for the family but small enough for the
> individual to run and care for. It will be constructed of
> the best materials, by the best men to be hired, after the
> simplest designs that modern engineering can devise.
> But it will be so low in price that no man making a good
> salary will be unable to own one – and enjoy with his
> family the blessing of hours of pleasure in God's great
> open spaces.[7]

The final line is interesting, because it completes the link between the cycling boom of the 1890s and growth of mass motoring. It suggests that Ford saw the greatest benefit of car ownership not as easier travel to work, or as closer ties to family and friends, or as widening the choice of places that people could live, but as granting access to the countryside for the purposes of recreation. His vision therefore mirrored exactly that of the early motoring elite, which in turn followed that of the late Victorian cycle fanatics before them: the wheeled vehicle was a way to escape the constrictions of the city and find freedom and joy in nature.

Ford's ambition required that small profit margins on each vehicle were offset by mass production, and this could only be achieved through his great transformation – the assembly line. Previous manufacturers had brought components to the car, but Ford's assembly lines took the car to its components. Each section of the ever-moving line was dedicated to a particular process: the painting was done at one part of the line, the engine inserted at another, the wheels added further along and so forth. This allowed each part of the car's creation to be optimised, undertaken by workers skilled and practised in that procedure with all the necessary materials and machinery close at hand. The speed of the lines, and the dependency of each stage on the previous, ensured that workers could not slacken off – the pace was relentless, even if the wages were good. When the process was perfected, a new car could be created from its hundreds of component parts, through thousands of different processes, in an hour and a half. Everything was geared towards producing the greatest number of cars in the shortest time and at the lowest possible price. Ford saw that his output would be maximised if he focused on producing a single model. His famous line 'Any customer can have a car painted any colour that he wants, so long as it is black' reflects this philosophy: having to make cars in different colours would complicate and slow down the manufacturing process, and black paint was the cheapest.

The most famous vehicle to roll off Ford's assembly lines was, of course, the Model T – the universal car. The first Model T, or 'Tin Lizzie', left the factory in October 1908 and the last one rolled off the US production lines in May 1927, some 15 million vehicles later (an average production rate of 2,200 cars per day). The Model T was not the fastest vehicle on the roads, but it was certainly the most numerous; as Ford himself said, 'There's no use trying to pass a Ford, because there's always another one just ahead.' It succeeded thanks to its great strength and ruggedness, which it gained from Ford's revolutionary use of vanadium steel for the chassis, its reliability, its ease of driving and, most important, its price, which plummeted from $900 in 1910 to $260 in 1925. So well made was the Model T, and so many sold, that more than 50,000 of them survive in working order today.

We have no record of what make of car conveyed Dayton and Lillian Stoner on their historic journey of roadkill discovery through rural Iowa in 1924, but at the time more than half of all the registered cars in the world were Fords. The Model T's cruising speed of 25 mph neatly matches the Stoners' own description of their rate of progress, so it is tempting to conclude that the world's first road ecologists were driving the world's first mass-produced car. We certainly know from his own journals that Joseph Grinnell's early observations of motorised ecocide were made from behind the wheel of a Ford.

The success of Ford's methods showed manufacturers on both sides of the Atlantic that there was an almost limitless market for small, inexpensive family cars, and they rushed to meet it. British rivals to the dominance of Ford started to appear in the 1920s: the Austin Seven (from 1923), the Triumph Seven (1927) and the Morris Minor (1928) were all hugely successful 'big cars in miniature'. Ford realised that his Model T was too large for the British market, and in 1931 his factory in Dagenham started to produce the Model Y, or Ford Eight, the first car the company designed exclusively for sale outside the USA and the first family saloon to sell for less than £100.[8]

The scale of these early operations was extraordinary, as it is today. A dramatic eyewitness account of an American car factory in the early 1930s is given in *Man and the Motor Car* (1936):

> Thousands of men are at work, and scores of different operations are under way all at once. Loaded freight trains are arriving; men are moving their cargoes to every part of the plant. Over in one corner you see what looks like a blaze of fireworks, it is merely a worker welding the top of a car. In another employees are building cushions, placing them when they are finished on a moving conveyor which carries them nearly a quarter of a mile away where they will be ready when the body of the car is assembled. On a similar conveyor, running along what is called 'the assembly line', the entire car is put together piece by piece. Men are stationed at regular intervals along this line, each one responsible for attaching some vital part to the moving car. All is carefully planned, for the evolving car must arrive at each station just when the worker responsible for attaching a wheel or a hub cap has finished with the car ahead. At one point an overhead hoist lifts an engine and swings it down in precisely the place where it belongs. Farther on another hoist that looks like a huge pair of ice tongs, places a completed body on the chassis. And so on, each man to his separate duty, until

> finally, at the end of the line, a worker steps into the car,
> starts the motor and drives off.

As cars became cheaper and more readily available, the balance of power began to shift in their favour. In cities, cars were promoted as a way of reducing the number of horses on the streets, thereby cleansing thoroughfares of the thick layer of 'mud' (Edwardian prudery for dung) that spattered over pedestrians whenever a carriage passed. Cars, unlike horses, did not require mountains of fodder to be imported into cities each day. Many former critics, some of them devout cyclists, swallowed their qualms and became converts to the cause of motoring. Fear and loathing of these new vehicles became dulled by exposure to them. The stone-throwing and the mockery started to dry up as falling prices fostered the aspirations of the stone-throwers and the mockers.

The car picked up momentum, and popular motoring shifted into second gear.

The forces of motordom

The growth of motoring in Britain can be charted through a list of firsts: the first number plates (1903), road tax (1909), petrol station (1919; Aldermaston), bypass (1923; Eltham, London), one-way street (1924; Hackney, London), traffic lights (1927; Wolverhampton), cat's eyes and pedestrian crossings (1934) and driving tests (1935). No new traffic-limiting measures were required during the Second World War or the subsequent period of austerity, as petrol rationing, the blackout and a shortage of parts and money led to a temporary reduction in motoring that would not occur again until the COVID-19 pandemic of 2020. But once post-war austerity was over, the restrictions picked up again to cope with the rapid rise of traffic: yellow no-parking lines (1956; Slough, Berkshire), parking meters (1958; Mayfair, London), motorway (1958; Preston), breathalyser (alcohol) tests and national 70 mph speed limit (both 1967), speed cameras (1982), compulsory use of seat belts by front seat occupants (1983), compulsory use of seat belts by all occupants (1991) and congestion charges (2003, London).

As the car's influence grew in the early years of mass motoring, so that of other road users diminished. In Britain, concerns were raised as early as 1903 that 'No other class of invention in this country has ever had to rely on conditions which sacrifice the public convenience to the privilege of the few'. Roads were originally places of commerce, relaxation and social intercourse as well as channels of movement. They were one of the few places in crowded cities where children could play. Although they were designed for movement in one of two directions, it was easy to move across them. But the unstoppable juggernaut of the booming car industry soon pushed pedestrians and other road users to the margins. Roads became conduits for cars but barriers to pedestrians.

During the 1920s, a decade that saw a quarter of a million Americans killed in traffic accidents, battle lines were drawn between different groups of road users. Worried by petitions to introduce speed limits in towns and keep some highways free of cars, the motor industry fought back with a range of dirty tricks that aimed to appropriate roads for the car once and for all. In his book *Fighting Traffic* (2011), Peter Norton tells how pressure from the motor industry and related interests (the 'forces of motordom', as he describes this aggressive alliance) subtly brought about changes in public perception. Car lobby groups started taking over school safety education, pushing the new mantra that 'streets are for cars and children need to stay off them'. Children forcibly evicted from their former playgrounds were then somewhat cynically recruited to hand out 'rebuke cards' to walkers who dared to cross the road anywhere other than at a few designated crossings. In the USA, the term 'jaywalker' (from those garrulous birds the blue jays) was contrived by motordom to ridicule recalcitrant pedestrians, painting them as country bumpkins who wandered across roads oblivious to the traffic around them. To add injury to insult, jaywalking was soon elevated from a term of mockery to a crime, a clear statement that the car had won possession of the streets in law.

Before 1920, the press generally placed the blame for accidents on drivers because the stories were fed to them by the victims or their families, but accounts of traffic accidents were increasingly fed to the newspapers, with appropriate spin and generous advertising revenue, by the motor industry itself; thus the finger of blame swung around towards pedestrians. In 1923, the *Chicago Tribune* carried an article by the Chicago Motor Club claiming that the 'reckless pedestrian' was responsible for 'almost 90%' of collisions between cars and people, and that jaywalking was to blame. At the same time the American motor industry developed strong messaging that road building was a public responsibility, and that objection to its expansion was an unpatriotic repudiation of progress. Cars gave people freedom, and freedom was enshrined in the Constitution; therefore driving a car was a constitutional right and impeding its progress an infringement of civil liberties.

Similar arguments had been proposed in Britain from the very start of motoring. In 1896, *The Autocar* magazine complained bitterly that 'innovation and enterprise are throttled at their birth in this land of so-called freedom', and a letter to the *Daily Mail* in 1903 railed against the undue influence of 'a few country folk who do not realise that the world is progressing'. A British motoring magazine article of the 1930s contrived, without any apparent shame, to paint the motorist as the saviour of the pedestrian: 'Nobody who drives a motor vehicle in the streets of London can fail to be astounded at the folly of which pedestrians are capable. The risks they take appal the man at the wheel who – it is no exaggeration to say – is constantly saving the life of walkers.' With such arguments, together with growing vehicle ownership and lower prices, the streets were soon won for the car. Mass motoring slipped from second gear to third, and now there was nothing to stop it.

The car has driven roughshod over any opposition, and the many arguments that have been levelled against it – whether on the grounds of health, safety, environment, aesthetics, equality or practicality – have all ended up as roadkill.

Our world has been utterly transformed by the car. There is barely a facet of our environment, natural or built, that has escaped its influence. Towns have been turned inside out, their commercial centres moved out to the fringes to allow easier parking for motorists. The architecture of our houses, public buildings and cities has changed to accommodate the housing and servicing of our vehicles, and acres of green space have been reduced to concrete and tarmac for them to sit idle on. Cars and their tarmacked habitats have stolen huge areas of cities that could be used for other purposes more beneficial to our economic and physical well-being: a car requires seventy times more city space than a cyclist or a pedestrian.

The architecture of our lives has similarly changed. Everywhere is now easily accessible from everywhere else, both a blessing and a curse. Thanks to the wider horizons offered by the car, people can now settle, work and raise their families further from the places they were born than their grandparents did. The Industrial Revolution sucked people from the countryside into the cities in search of work; generations later, the car allowed them to return.

Cars slip easily through the social and economic membranes that separate urban from rural, bringing benefits and problems in equal measure. Roads inject people into wild places, and thereby tame them. Many rural regions have experienced a period of prosperity and growth they would not have enjoyed had cars not made them accessible to people, money and ideas from the cities. But this has also led to the erosion of regional characteristics and idiosyncrasies, to the infiltration and dilution of centuries-old communities. Nowhere is unchanged, and everywhere regresses towards a lacklustre average: our social and cultural landscapes have been homogenised by the internal combustion engine. Laurie Lee described its impact on one English village: 'Then, to the scream of the horse, the change began. The brass-lamped motor car came coughing up the road. Soon the village would break, dissolve and scatter, become no more than a place for pensioners.'

We are prepared to forgive the car some appalling crimes against humanity. Each year, millions of people around the world pay for its convenience with life and limb. The World Health Organisation (WHO) estimates that nearly 1.5 million people die in road traffic accidents each year, more than are killed by malaria, and that between 20 and 50 million more are injured. Road accidents are the world's leading cause of death among people aged between 5 and 30 years. Outdoor air pollution, much of it from traffic emissions, kills over 4 million more (some estimates put it as high as 9 million). The number killed by noise pollution from traffic each year is unknown, but based on data from Western Europe it is likely to run to hundreds of thousands. Road traffic brings a global pandemic of death and injury that no government seems willing to lock down.

Of course, cars and other forms of motorised transport also bring huge benefits, an argument that is so self-evident that it hardly needs making. Why, otherwise, would most of us (me included) go to the expense and trouble of owning one, and why would we be so prepared to put our lives at risk each time we get behind the wheel? Like the Stoners in the 1920s, we all know that driving is bad for the environment, and potentially fatal to ourselves, yet we still do it with barely a second thought. Driving incurs huge costs on the user, yet it somehow prevails over them. Most people who suffer serious accidents on the roads, or who lose family or friends in the same way, sooner or later get back behind the wheel.

The first component of traffication

If there were a general term to describe the expansion of road networks and the burgeoning of motorised travel along them – very strangely there isn't, so let's call it 'traffication' – it would encompass a number of interconnected but qualitatively different processes. Three of these are central to the story that will unfold. First, cars and other motor vehicles have become more numerous; second, they have become faster; and third, they have become more pervasive in the environment.

The first and most manifest element of traffication has been the growth in vehicle numbers. The rise in traffic volume on Britain's roads has been relentless; apart from a slight wartime drop between 1939 and 1946, the number of motor vehicles has risen each year since records began in 1909. There are now close to 40 million registered motor vehicles in Britain, over three-quarters of them cars. But traffication cannot be measured simply by the number of vehicles on the roads; more important is how much driving we do in them. This is where the concept of the 'vehicle mile' comes in helpful, combining into one measure the number of vehicles on the roads and the distance that each of them travels. Thus one vehicle travelling 1,000 miles, or two vehicles each travelling 500 miles, or ten vehicles each travelling 100 miles, all generate 1,000 vehicle miles. Figure 1 shows the number of vehicle miles travelled in Britain each year between 1950 and 2020. Inevitably, the pattern has been one of steady growth (with one unprecedented exception); the oil crises of 1973 and 1979 did little to slow this long-term increase in traffic volume.

These increases have been so gradual, a rise in traffic volume of 1 or 2 per cent each year, that most of us have barely noticed them, but the cumulative effect across a human lifetime has been profound. In the spring of 1947 the ornithologist David Snow, then a young undergraduate in Oxford, cycled with a friend out to Faringdon to look for a colony of breeding herons that they had heard about. In his autobiography *Birds in our Life*, published some sixty years later, Snow wistfully recalled the end of that day:

> It took longer to find than we had expected, and when
> we were ready to return it was almost dark. We had no
> bicycle lamps and decided that we would ride, but that
> if a car came along we would get off and walk until it
> had passed. In the course of the 17-mile ride along the
> important Oxford–Swindon main road, we had to get
> off once.

That road is now the busy A420, and anyone attempting to repeat Snow's journey under the same conditions today would find themselves pushing their bicycle for 17 miles.

Since the launch of the first Space Shuttle and the introduction of the mobile phone in the early 1980s, the volume of traffic on our roads has more than doubled. It was not until COVID-19 hit in 2020 that any sizeable drop in road travel was recorded, returning us briefly to the traffic levels of the late 1990s.

The skin of our planet is crawling with metal lice. Every square mile of England's once green and pleasant land is being battered by an average of

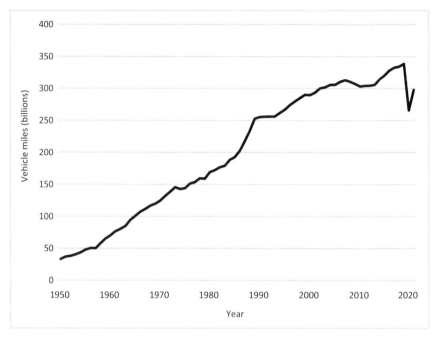

Figure 1. The inexorable rise in traffic volume in Britain, measured in billions of vehicle miles driven each year, between 1950 and 2021 (cars, goods vehicles and motorbikes combined). The sudden drop in 2020 and 2021 was due, of course, to the COVID-19 pandemic. Data from the UK Department of Transport.

6 million vehicle miles each year; that's equivalent to ten cars being driven at 70 mph, day and night without cease, around each square mile. Britain's cars alone would form a metal belt three vehicles thick around the entire length of the Earth's equator. The total number of cars on the planet is estimated at somewhere between 1.2 and 1.5 billion, together with another 400 to 500 million motorised vehicles of other types. Taking just the cars, that's a bumper-to-bumper traffic jam more than 5 million km long – fourteen parallel lines of stationary vehicles stretching from here to the Moon. More will be added as countries such as China, India, Brazil and Russia race to catch up with the levels of car ownership currently seen in North America, Western Europe and Japan. Some experts predict that by 2040 there will be 2 billion cars on the world's roads and perhaps 800 million other motor vehicles.

The second component of traffication

The second component of traffication is speed. The British Motor Museum, not far from the country's historical vehicle manufacturing centre of Coventry, is a great concrete cathedral raised in veneration of the car. Its vast

The extraordinary speed of car development, and the extraordinary development of car speed, is illustrated by these exhibits in the British Motor Museum. Just half a century separates Karl Benz's original motorised tricycle (1886; this one is a replica) from the sleek MG speed record cars (1938 and later) in the background. The Benz had a top speed of 9 mph, the 1938 MG over 200 mph.

amphitheatre houses hundreds of largely British-made vehicles. If the main exhibition leaves the visitor hungry for more, an adjacent building houses the equally impressive reserve collection. Here, cars are parked nose-to-tail in roughly chronological order, horseless carriages at the front and SUVs at the back, like a traffic jam that started backing up at the dawn of motoring.

Each vehicle in the museum has an informative label that describes its historical importance and presents information on its year of production, its sale price when new and its top speed. In the summer of 2021, I spent a happy day wandering around the collections, jotting down some of these numbers in my notebook. I tried to include only cars built for the mass market, those likely to be representative of the majority of vehicles on the roads at the time, and ignored those built for racing or the luxury market. This being a museum, there were few 'ordinary' cars in the collection that were built after 1980. To find out what has happened to the top speed of the average car since then, I used industry lists to identify the ten top-selling cars in the UK in each of five years between 1985 and 2021 and looked up their top speeds in online catalogues. The results of this little bit of impromptu research are shown in Figures 2 and 3 (pp. 32–33).

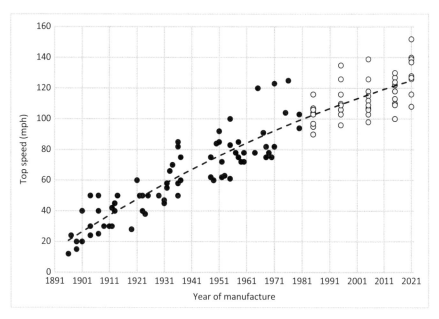

Figure 2. The increasing top speed of 'ordinary' British cars, 1895–2021. The black dots show data on cars (mostly British-built) in the National Motor Museum; the grey dots show the ten best-selling cars (of any make) in the UK in each of five years since 1980. Note the absence of new vehicles during the war years of 1939–45, when most car factories were turned over to armaments production.

Figure 2 shows that cars have become faster. This will come as a surprise to absolutely nobody, but as with most graphs there is something of interest to be gleaned if you look at more closely. The first thing to notice is that the entire industry has moved gradually in the same direction: none of the pre-1930 cars in my sample had a top speed above 60 mph, and none of the cars produced after the Second Word War had a top speed below 60 mph. There are no outliers – no cars in the top left or bottom right sectors of the graph. The second point made by the graph is that growth in speed has been steady: there have been no sudden upward leaps to mark a seismic change in technology or taste. The third thing the graph tells us is that while speeds have continued to increase, they have done so at a slowing rate: the dotted line that marks the average is not straight but levels out as time has gone on. Top speeds doubled in the two decades between 1900 and 1920, but it took half a century for them to double again (1920–70), and at the current rate of increase we won't see a doubling of 1970's average top speed of 90 mph for another century or more – in petrol-powered cars at least. This slowdown is perhaps unsurprising in a country in which the highest permitted speed on any road is 70 mph; we have to go right back to the

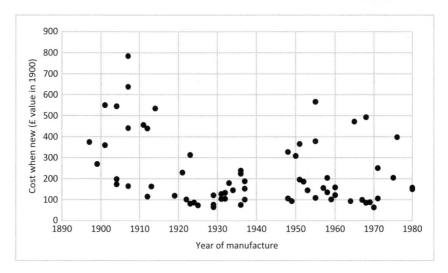

Figure 3. Price when new of 'average' British-built cars, 1890–1980, converted to their 1900 values. Again, note the absence of new vehicles produced during the war years of 1939–45.

1950s to find any car in my sample that was not capable of breaking today's national maximum speed limit.

Less predictable, perhaps, is the story told by Figure 3. This shows changes in the purchase price of my sample of 'ordinary' cars from the British Motor Museum, but I have adjusted them all to their 1900 equivalents by taking changes in money values into account. Here there is no very clear pattern, except perhaps that the average price of a new car tended to fall between 1900 and the outbreak of the Second World War.

Putting these two graphs together, the story that emerges is that a lot of extra speed has been given to motorists over the last century more or less free of charge, paid for by improvements in technology, competition and the economies of mass production. Recall that pioneering road ecologists Dayton and Lillian Stoner, in their 1925 article on roadkill, considered that a speed of 35 mph made it difficult or impossible for animals feeding on roads, or crossing over them, to escape. According to my graph, the top speed of the average British car in that same year was already around 50 mph, and it would have been higher still in American cars, which were usually more powerful than their British equivalents.

The flood of roadkill studies that appeared from the mid-1920s onwards was probably not simply a case of people reading the Stoners' original article in 1925 and deciding to jump on the bandwagon. Instead, I suspect that roadkill appeared rather suddenly as a new and conspicuous feature of road travel at around this time, as a critical number of cars started to exceed that 35 mph limit. This is borne out by Joseph Grinnell's observation in 1920 that roadkill 'is a relatively new source of fatality'. The fact that the bird-loving British did not start talking about roadkill until a decade later may have been because of our less powerful cars and narrower, more winding roads, which combined to keep speeds lower for longer.

Of course cars are not often pushed to their maximum speed, but having more power under the pedal allows drivers to reach their chosen cruising speed more quickly and tempts them to exceed it. As cars have become more powerful, so too the quality of road surfaces has improved, allowing vehicles to gain speed more quickly still. Even the Stoners may have succumbed to this temptation. From what I can glean from their articles, the couple covered an average of 160 miles per day during their original journey in 1925, compared with 270 miles per day during a trip they made in 1935 (of which more later). If we assume that they drove for eight hours each day, including stops, then it seems likely that during their 1935 trip they would, at least occasionally, have exceeded the 35 mph they had so decried in their article of a decade earlier. The staid and stolid Stoners had become, in their own word, 'speeders'.

Data from the UK's Department of Transport suggests that the average speed of vehicles on Britain's roads exceeds the maximum legal limit for the respective type of road. This is the *average* speed, and since many drivers travel at or below the limit, it follows that many others exceed it by a significant margin. Around half of all car drivers in Britain, and well over half of all motorcyclists, regularly exceed the speed limit, particularly on motorways and residential roads. Research has shown that drivers strike a compromise between their desire for higher speed and the risks of detection by the police and of accident to themselves (although not necessarily to others, and certainly not to wildlife); in other words, we tend to drive as fast as we consider it selfishly safe to do so.

As a car's speed increases, so it produces more pollutants – exhaust gases, noise, micro-particles from tyres and brake linings. It also takes longer to stop should something unforeseen appear on the road ahead, and it gives that unforeseen something less time to get out of the way.

The third component of traffication

The third component of traffication is prevalence. The number of cars on our roads and the speeds at which they travel have both increased hugely over time, but they could, in theory at least, have done so in such a way that left large areas of the country unaffected. Had this huge growth in traffic been restricted to certain regions, for example, or to certain types of roads, then large parts of our countryside might have escaped their influence. But that is not what has happened: national statistics clearly show that traffic volume has increased almost uniformly in different regions, whether urban or rural, and on all classes of road. Furthermore, over 60,000 miles of new roads have been added to Britain's already dense network since 1950, and many older highways have been widened to accommodate more lanes. The road network has expanded, and the area of our countryside that remains unaffected has correspondingly shrunk. Although the USA is generally regarded as the world's car capital, Britain's road density (the average length of road per square km of land) is more than twice as high.

One way to examine the issue of prevalence is to look at how much of the country's land lies within different bands of distance from the nearest road. Maps 1–4 (pp. 36–39) show the area of land in Britain that falls within four bands of distance from one or more roads. Because detail can be lost at this scale, part of central England is shown at a higher magnification.

The first map shows all the land that falls within 100 m (around the length of a football pitch) of one or more roads. The individual roads are clearly visible, and at this scale the network appears as thin as a spider's web, but this is deceptive – already a fifth of Britain's entire area, and nearly a third of road-rich England's land, is shaded.

The second map shows all the land falling within 500 m of one or more roads. Now the details of the road system start to blur, as spaces between adjacent roads become filled in. Well over half of Britain is less than 500 m of a road, and in England three-quarters of the land is now shaded. If you are a hunting barn owl, or a wandering hedgehog, or a pollen-laden bee returning to the hive, sudden death is never very far away in this zone.

Map 3 shows the area of land falling within 1 km of one or more roads. Now the lines of road disturbance have coalesced and cover vast swathes of the country; over much of the south and east of the country only small windows of calm show through. Nearly 80 per cent of Britain's land is now shaded, and over 90 per cent of England, yet nowhere in this huge area are you more than a few minutes' walk from a road.

The final map shows all the land falling within 2 km of a road. South of a line from the Severn to the Humber, the only remaining car-free windows of any size are in the Broads and the Brecks of East Anglia, on Salisbury Plain (much of it reserved as a military training area) and on the high moors of Devon and Cornwall. North of the Severn–Humber line, the hills and mountains of Wales, northern England and Scotland remain largely clear, but most lowland areas are shaded.

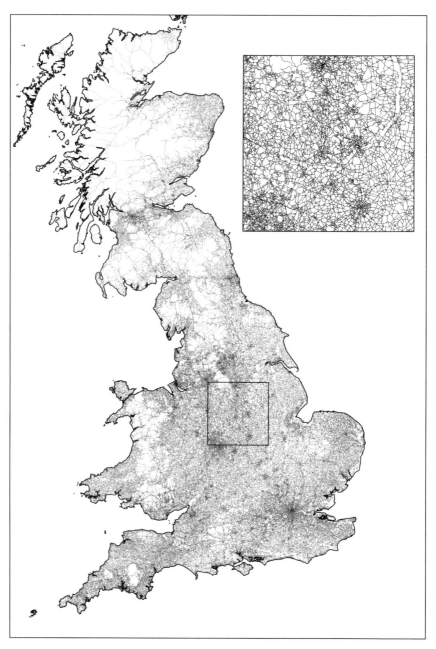

Map 1. The areas of Britain falling within 100 m of a road, shaded in black. The inset is a magnification of the square in central England. This map and the following three slightly underestimate the pervasiveness of road traffic for two reasons: first, over 100,000 km of local access roads and unpaved tracks are excluded; second, distances are measured from the middle of each road, not from its edge.

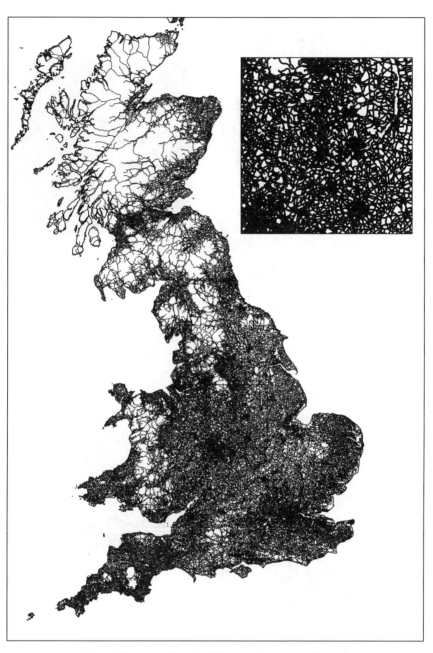

Map 2. The areas of Britain falling within 500 m of a road.
See caption to Map 1 for further details.

Map 3. The areas of Britain falling within 1 km of a road.
See caption to Map 1 for further details.

Map 4. The areas of Britain falling within 2 km of a road.
See caption to Map 1 for further details.

These maps show just how completely road traffic has infiltrated our landscapes and our lives. Britain is one of the most pervasively trafficated countries in the world. The 80 per cent or so of our land that falls within 1 km of a road is well above the comparable value measured across the whole of Europe, which is just under 60 per cent.

Most of us live near roads, and most of our wildlife does too. We are in their thrall. Anything produced by traffic that spreads up to 500 m from the road – air pollution, perhaps – affects 60 per cent of Britain's land and 75 per cent of England's. If we assume that, on a calm day, the sound of vehicles can be heard up to a kilometre away (it's often much further, but we will return to this later), then traffic noise pollutes nearly 80 per cent of Britain's land, and less than 10 per cent of England is free of its roar.

The country's people, and its wildlife, live in ever-closer proximity to increasingly busy roads carrying faster vehicles – the unholy trinity of traffication.

Before the arrival of COVID-19, Britain's drivers clocked up around 350 billion miles of road travel each year – a trip to the Sun and back every four hours. Other nations travel much further. Since the crew of Apollo 11's lunar module stepped onto the surface of the Moon in 1969, their compatriots have driven to the next star system and back – twice. Now add in the miles driven each year in China, Japan, Brazil, India and all the world's other road-rich nations, and the distances become unimaginable. In our cars and buses and trucks we lap the deep-space orbit of Halley's Comet five times every single day. The amount of driving we do is quite literally astronomical.[9]

But of course these sci-fi voyages are not made through the sterile silence of space. They are somehow compressed onto the dry third of a tiny blue-and-green-and-yellow-and-white planet. And they run around, through and often directly over the only life we know in the universe.

CHAPTER 3

'An Inconspicuous Splotch of Red'

T en years after the publication of their groundbreaking article 'The Toll of the Automobile', the first ever to document the car's fearsome capacity to flatten wildlife, the Stoners were on the road again. This time, Dayton and Lillian documented the roadkill they encountered during a journey between Albany, New York, and their old home in Iowa City. Their round trip of over 2,000 miles took the couple through eight states in eight days. The following year (1936) they published their results in an article entitled 'Wildlife Casualties on the Highways', presenting their readers with yet another litany of death: five hundred flattened little bundles of scales, skin, feathers and fur lying forlornly on the road. The species they recorded most frequently this time were the house sparrow, common skunk and cottontail rabbit, as well as plenty of chickens and a distressingly large number of pets (over thirty cats and two dogs). Most of the snakes they found were entered on their dead-list as 'miscellaneous', presumably because they were too mangled to allow them to be identified to a particular species. Intriguingly, they did not record a single red-headed woodpecker, the most frequently logged victim during their trip a decade earlier.

Their total included only two frogs, and speculating on this low casualty rate led the Stoners to another important insight: the victims observed through the windscreen, they realised, represent only the tip of an iceberg of roadkill. Animals such as frogs and toads are greatly under-recorded as casualties because, in Dayton's words, 'their bodies are soft and yielding, so that even immediately following death by motor traffic, little of the animal remains except an inconspicuous splotch of red or a moist spot in the highway'. We know now that the numbers of amphibians killed on the roads exceed those of less 'soft and yielding' animals such as mammals and birds by a factor of tens or even hundreds.

As interest in the study of roadkill grew in inter-war USA, and later spread to Europe and beyond, ecologists attempted to refine their hitherto rather haphazard observations. Thomas G. Scott, another Iowan, recognised that many victims of roadkill are quickly removed by predators or scavengers, so counts of the little bodies that remain on the road greatly underestimate the true death toll. In 1938, he tried to build this into his estimates of total roadkill by assuming that the bodies of victims would be visible to passing motorists for four days after they were struck (we know now that they usually disappear far more quickly). Scott also tried to bring a spatial element to the study of roadkill by noting how far from his starting point each corpse was seen, so that he could later plot the carnage on a map. As the science grew, its practitioners increasingly sought to link their observations of roadkill (or, as scientists prefer to call it these days, WVC: wildlife–vehicle collision) to features in the surrounding landscape, weather conditions, traffic flow and so forth. Dead-lists have not ceased – many are still published each year – but roadkill research has matured over the years into a science that goes beyond the presentation of simple death tolls and instead tries to explain where, when and why animals meet their end on the tarmac.

The first law of roadkill

The many hundreds of studies that followed the Stoners' original article of 1925 have, over time, coalesced around a single, unifying truth of roadkill, one that forms a central pillar of the branch of science we now call road ecology. To understand this one insight is to understand the essence of roadkill. It is deceptively simple and can be summarised in just six words: *roadkill is not a random event*. Every flattened hedgehog you see, every eviscerated fox or broken barn owl, is a casualty not of haphazard chance but of cold, harsh mathematical probability. Next week's roadkill can be foretold in a way that next week's lottery numbers cannot.

The theory that underpins this fundamental insight can be illustrated with a simple example. Let's take a badger that crosses a moderately busy road twice each night – once on its way out to forage after darkness falls, and again on its way back to the sett just before dawn – and assume that its risk of being killed each time it crosses that road is one in a thousand. We can combine these two pieces of information – the number of crossings and the death rate per crossing – to predict that our badger stands about a 52 per cent chance of becoming roadkill within a year, and that its odds of successfully dodging the traffic for two years are less than one in four.[10]

In this made-up example, neither the number of crossings that the badger makes each night nor its chances of being killed each time seem unrealistically high. Badgers are often found in areas with high densities of busy roads and they can travel far each night. And if a roadkill risk of one in a thousand (or 0.1 per cent) per crossing sounds high, then pity the poor amphibians: a Danish study estimated that the roadkill risk of frogs and toads on busy roads is between 34 per cent and 98 per cent *per crossing*. Similar studies from Germany and the Netherlands have found that even with a traffic volume of less than one vehicle per minute, up to 90 per cent of crossing frogs and toads are flattened. Few amphibians survive to cross a busy road twice. Reptiles do not fare much better: black ratsnakes in Canada suffer a 3 per cent risk per crossing, thirty times higher than that of our imaginary badger, and a study of snakes attempting to cross US Highway 441 found that most of them were killed in the first lane they tried to cross; the few that made it as far as the second lane all died there. Freshwater turtles trying to cross US Highway 27 in Florida suffer at least a 98 per cent mortality rate. Our imaginary badger's collision risk of 0.1 per cent per crossing may therefore be rather on the low side, yet even with such modest levels of risk the vicious mathematics of cumulative probability dictate that our badger is quite likely to end up as roadkill.

The same statistical qualities of probability also show that small changes in behaviour can make enormous differences to an animal's chances of survival. If our badger decides, just one night each week, to visit a different feeding site, requiring it to cross a road that is five times more dangerous than the first, then its chances of becoming roadkill within a year soar from around half to more than two thirds.[11] And we can use the same methods to predict roadkill from a different perspective, that of the driver. If our badger is one of ten animals that cross the same one-in-a-thousand risk road twice a night, then we can work out that our chance of passing a fresh corpse as we drive by on any particular morning is around one in fifty.

So if we know how often an animal crosses a road, and what its chances of surviving each crossing are, then we can predict how likely it is to end up as roadkill. The problem is that roadkill in the real world is very much more complex than the example of our badger suggests, because both components of the overall roadkill probability are hugely variable and dependent on many complex factors.

Let's take road-crossing rates first. A real-world badger will forage in different places each night, and may move further to feed at some times of year than at others, meaning that the number of roads it crosses will constantly vary. Its road-crossing behaviour will certainly change over the course of its life. A young, inexperienced badger is particularly vulnerable as it leaves its birthplace and wanders widely across a strange and dangerous landscape in search of a new territory. We know that roadkill rates are particularly high at this period of most animals' lives. If it survives its perilous youth, then perhaps the experience of a near miss or two with cars might cause our badger to become warier of crossing roads in adulthood. Later still in life, its road-crossing behaviour might change again, if for example it is driven out of its territory by a younger rival and forced into a lifetime of hazardous wandering. Throughout its life, fluctuations in food availability, weather, disturbance from people and a whole host of other factors will cause it to change its road-crossing behaviour on an almost nightly basis.

The second component of roadkill – the chance of our badger making it over safely once it has decided to cross – is similarly dependent on a complex range of factors. The volume and speed of traffic when the crossing is attempted, each driver's ability to see the badger and their willingness or ability to take evasive action will all play a big part in determining whether or not the animal makes it over to the other side. Each of these in turn will vary with the time of night, the month of the year, the day of the week, the weather, the extent of street lighting and a host of other factors. Even the phase of the moon, which determines the amount of light falling on unlit roads at night, has been shown to affect roadkill rates. Each individual badger will also vary innately in its roadkill risk, according to its age and sex, its previous experiences of crossing roads, its health and the behaviour it adopts (freeze or flee) when it detects danger.

To add to all this complexity, the number of road crossings our badger attempts and the risk it faces each time are themselves interrelated – the higher the risk, the fewer the crossings it will make. Badgers might cross quiet roads, with a low roadkill risk, without a second thought, and might occasionally decide to risk crossing busier and more dangerous highways, but they will usually baulk at crossing a busy motorway, which carries the highest per-crossing risk of all. Indeed, we know from studies of badgers and other mammals fitted with tracking devices that this is exactly what happens – they usually stay well away from the busiest roads. So it could be that quieter, apparently less dangerous roads actually pose the highest overall risk of sudden death, because the low risk per crossing is more than offset by a large number of crossings made.

Many animals are aware that roads are dangerous places, and they cross busy roads far less often than would be expected given their normal patterns of wandering. Electronic sensors have shown that the heartbeat rate of American black bears rises as they approach a road, and it gets faster the more traffic there is. Bears clearly sense roads as a danger, and they become nervous. Because of this, they tend to cross busy roads at night, when there are fewer vehicles around; grizzly bears and European brown bears behave in exactly the same way, and are most frequently killed by traffic around midnight. Other animals, such as some insects, amphibians and reptiles, appear to have little if any awareness of the dangers posed by traffic, and often cross busy roads as readily as quiet ones, suffering very high roadkill rates in the process.

An animal's risk of roadkill also depends on its own particular ecology. A bat may cross the same busy road many times in a single night, especially if it is feeding on insects attracted to streetlights, but if those crossings are made above the height of the average car then its chances of sudden death will be low. If an industrial estate opens nearby, however, and taller goods vehicles start using the road at night, the risk to hunting bats might suddenly increase. Summer rainfall is important in determining whether a hedgehog survives the traffic, because it affects the number of slugs and snails available for it to feed on, and hence the distance it needs to walk between meals. Hibernating mammals often suffer high rates of roadkill in the weeks before they bed down for the coming winter, as they travel further to find the extra food they need to sustain themselves during the long sleep. Birds such as blackbirds and song thrushes may be relatively safe from traffic for most of the year, but if they nest in roadside hedgerows they become acutely vulnerable, swooping low over the tarmac with beakfuls of food for the chicks every few minutes, oblivious to everything but their parental obligations. If the parents are killed, their dependent chicks become collateral roadkill victims too, starving to death without leaving the nest.

Roadkill, then, is not simply the random convergence of a car and an animal at a particular time and place. It is a process that can be described using mathematics, allowing us to foresee that some species, and some individual animals, are more vulnerable than others, and that some roads are more dangerous. But just because something is predictable in theory does not necessarily mean that we are able to predict it very well in practice. Roadkill is an immensely complicated phenomenon, and our understanding of its many ecological, behavioural, sociological and physical nuances remains poor. We know too little about how animals move around the landscape, how they decide whether or not to cross a road or what their chances of crossing safely are to be able to predict with any great accuracy where, when and how roadkill will occur. We are not much better informed about the human side of the roadkill equation – there is very little data available on traffic speed and volume, or how it varies in time and space, and we know as little about the behaviour of drivers towards animals on the road as we know about the behaviour of those animals towards cars.

Nevertheless, the fundamental insight that roadkill is governed by the laws of mathematical probability, rather than by the ungovernable vagaries of random chance, is a hugely important one because it enables us to start breaking the problem down into its component parts. This in turn allows us to start developing measures to reduce them, and thereby lower the overall risk.

Patterns in roadkill

We may understand rather little about the statistical undercurrents that dictate where and when roadkill occurs, but we are not entirely ignorant of them. From its very first days, the dead-list mania unleashed on the world in 1925 by Dayton and Lillian Stoner started to identify some broad and consistent patterns. In their very first published offering on the subject, the Stoners recognised that road surface and traffic speed are important determinants of roadkill – they found more dead animals on better-surfaced roads, which allow cars to go faster. This is perhaps the most consistently reported finding in roadkill research; if the first law of roadkill is that it is not a random process, then its second law is surely that casualty rates are low where traffic is slow.

A number of other equally consistent but perhaps less obvious patterns have also emerged. The earliest British contribution that I can find on the subject of roadkill is an article published in 1936 entitled 'The Death-Roll of Birds on our Roads', in which a certain Maurice D. Barnes (who seems to have left no other trace in the history of ornithology) described the results of his cycling census of road-killed birds. The resolute Mr Barnes covered 4,000 miles on his bicycle in 1935, 132 of them in a single day, logging 940 dead birds along the way. Rather oddly, given all this effort, he omitted to tell his readers where he went or how many dead birds of each species he saw, merely stating that the most frequent victims he encountered were 'Sparrows, Finches, Yellow-hammers, Thrushes, Blackbirds and Larks'; perhaps he was a better cyclist than an ornithologist. But just like Joseph Grinnell and the Stoners, a decade before and a continent away, Barnes could clearly see the wider implications of his observations: 'If this carnage is going on throughout the country, and there is every reason to believe this is so, then the annual death-roll must be simply enormous.'

Barnes made two important observations. First, he found that casualty rates are highest where hedges line the road, and particularly where they immediately border the highway. This, he suggested, is because birds flying low across the road from one hedge to another have little chance to react should a car pass between them. The importance of the type of habitats on either side of the road, and how close they come to the edge of the tarmac, has been borne out by many subsequent studies of roadkill around the world.

Second, Barnes showed that the rate of avian roadkill is not constant across the year, but peaks in the middle of summer and falls to a low in winter. This, he thought, was due to the popularity of summer motoring, the presence of large numbers of young birds still developing their flying skills and an abundance of insects on the warm roads. Again, seasonal fluctuations in roadkill rates have been reported by many subsequent studies, in Britain and elsewhere.

British dead-listing came of age in 1960, with the launch by that wonderful organisation the British Trust for Ornithology (BTO) of the first nationally coordinated survey of road-killed birds ever undertaken in any country. The results of their *Road Deaths Enquiry* clearly confirmed the seasonal pattern of roadkill in birds that previous writers, such as the cycling Mr Barnes, had already observed; roadkill rates of birds show a pronounced peak in the summer months and fall to much lower levels in winter (see Figure 4). But not all species show the same seasonal trends.

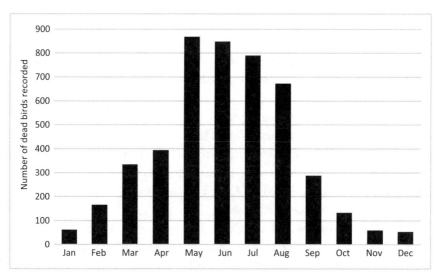

Figure 4. The first ever nationwide assessment of road-killed birds was undertaken in Britain in 1960–1 by the British Trust for Ornithology. It revealed a very clear seasonal pattern; this graph shows the number of road-killed birds (of the ten most frequently recorded species) recorded by survey participants in each month. The survey encouraged its participants to visit the same stretches of road regularly throughout the year, so the summer peak is not simply the result of observers going out more in nice weather.

For nocturnal species, this pattern may be reversed. Owls often suffer their highest roadkill rates in the autumn and winter months, when the period of darkness they inhabit expands to embrace the deadly evening and morning rush hours. A recent survey of roadkill in Britain, based on more than 50,000 observations sent in by members of the public, showed that many non-hibernating mammals, such as polecats, badgers and hares, show two peaks of roadkill, one in spring and another in autumn. The spring peak might be the result of animals moving around to find breeding territories, while the autumn peak perhaps reflects the naivety of their offspring as they spread out from their birthplace into an unfamiliar and dangerous land.

We know, too, that roadkill rates differ greatly between different species – some are inherently more vulnerable than others. Birds that tend to live solitary lives, with only one pair of eyes to spot approaching danger, are more susceptible to being struck than birds living in wary flocks. Larger birds appear to be more susceptible to roadkill than smaller ones, perhaps because they are slower to take off and less manoeuvrable. Some animals may be attracted to roads and spend significant periods of time on them, hugely increasing their chances of being struck by traffic. Reptiles, for example, often use roads as a source of heat to raise their body temperature, and nocturnal birds such as nightjars sit on them at night and hunt the insects similarly attracted by the warmth the tarmac radiates. Deer and moose may be attracted to roads to lick the de-icing salt we scatter on them in winter. Large numbers of waxwings have been killed flying across roads to feed on fruiting trees planted in the central reservation. Scavengers, from kites and foxes to ants and beetles, visit roads to feed on roadkill, often ending up in the same mangled state as their intended meal. Barn owls and other predators hunt along grassy roadside verges, making them particularly vulnerable. Seed-eating birds such as house sparrows visit roads to collect grit to help them grind down their food in the muscular avian pre-stomach known as a gizzard. This might explain why, in most lists of roadkill from Europe and North America, the house sparrow is one of the most commonly recorded casualties.

Smartness also appears to play a big part: a study from Denmark suggested that birds killed by traffic tend to have smaller brains, relative to their body size, than those that die from other causes. The same pattern was not apparent for other organs such as the lungs or liver.

A final consistent pattern that emerges from the hundreds of published studies of roadkill might seem to be such an obvious one that it is barely worth mentioning: it is that the species most frequently found dead on a road are generally those that occur in greatest numbers around that road.

Obvious indeed, but the rather more interesting question then arises of *why* these are the most common species around roads. There are some hidden depths to this pattern that we will return to in Chapter 10.

The toll road

We can, therefore, identify many broad patterns in roadkill. We know that some species, and some individual animals within those species, are more likely to end up as roadkill than others, and that their vulnerability changes in many complex ways. We know that some roads are more dangerous than others, and we have a good idea which those might be.

But we are still very far from being able to say, with any degree of confidence at all, how many animals are actually killed on our roads each year. Counting the roadkill that we see as we walk, cycle or drive along is relatively easy, but using those counts to work out the total casualty rate is not. As the Stoners realised, some victims leave more conspicuous and persistent remains on the tarmac than others. Frogs and other amphibians are more or less obliterated when they are hit; the *toad*kill that we don't see may greatly exceed the roadkill that we do. Predators and scavengers often devour the smaller roadkill victims or move them to safer places for eating; indeed, some birds such as crows and kites appear to have become particularly specialised in this form of feeding. And even the most sharp-eyed road-kill-counter will fail to spot some victims, particularly those that have been knocked into the verge. Mortally injured animals stagger, writhe or flutter away from the road to die, unseen and unlamented, elsewhere: research from Canada shows that for every large mammal seen dead on the road, two others are fatally wounded and missed from roadkill counts. The victims we see lying on the tarmac represent just a small fraction of the true body count.

Researchers have tried to examine some of these issues experimentally, for example by placing a known number of carcasses along a road and then revisiting the same stretch later to see how many of them remain. The results are startling – one study suggests that the true death toll for amphibians might be up to 40 times higher than the numbers counted dead on the road would suggest. Other studies have found that the bodies of nearly three-quarters of all the frogs killed by traffic are removed by scavengers within just a few hours, and that almost all dead birds and snakes are consumed or removed within a day and a half. The car kills, the scavenger takes away and the roadkill researcher underestimates the carnage.

Given all these difficulties and uncertainties, working out how many animals are killed on the roads each year is an almost impossible task, though

Toadkill. The larger animals we see dead on the road are just the tip of the roadkill iceberg. Smaller, softer-bodied animals such as frogs and toads are killed in far greater numbers than deer, badgers and hedgehogs, but their carcasses rarely persist for long on the road so their roadkill rates are greatly underestimated. *(Wikimedia Commons)*

that has not stopped a few brave souls from trying. In his article of 1935, Dayton Stoner combined his own estimates of the number of vertebrates killed each day per mile of road with those of several of his contemporaries. His back-of-the-envelope calculations suggested that 'an average daily motorcar casualty list of something more than 200 vertebrates per 1,000 miles of main highway is a fair approximation of the true conditions'. The fastidious Dayton questioned 'whether we are justified in applying the current findings of vertebrate death rates on a per diem basis to the 750,000 miles of improved roads throughout the United States', leaving to the less scrupulous reader the simple task of extrapolating his estimate to 150,000 vertebrates per day, or around 55 million per year.[12] But this calculation makes no attempt to account for the under-recording of the more 'soft and yielding' members of the USA's vertebrate fauna, nor does it take into consideration the removal of corpses by predators or scavengers. Roadkill on the many thousands of miles of unimproved highway was ignored in Dayton's calculations, and invertebrate victims were not even considered. These figures must therefore massively underestimate the scale of roadkill in 1930s America.

Another rough calculation made in 1998, not much more sophisticated than Dayton Stoner's estimates of more than 60 years earlier, suggested that around 360 million birds, reptiles, amphibians and mammals die on the USA's roads each year. A more recent study, published in 2014, suggested that the number of birds killed on the USA's roads each year lies somewhere between 89 and 340 million. In Canada, a country of around the same area as the USA but with less than a tenth the number of road vehicles, the death toll of birds has been estimated at around 14 million a year.

Across Europe, perhaps 200 million birds and 30 million mammals are killed annually on roads, although again there is huge uncertainty around these estimates. Small mammals such as rodents must, like amphibians, be greatly under-recorded in roadkill surveys. A survey in Sweden asked drivers to count the number of larger mammals they hit during a given period and to keep a record of the distance they had driven. By scaling up the responses to the total distance covered by all drivers in the country, the researchers estimated that around 13,500 moose, 59,000 roe deer, 81,500 hares, 33,000 badgers and 12,500 red foxes were killed on Sweden's roads in 1992 alone.

In Britain, perhaps 100,000 foxes, between 167,000 and 335,000 hedgehogs, 50,000 badgers and 74,000 deer are killed on the roads each year, and maybe something in the region of 30 million birds. But once again these numbers are little more than educated guesswork, and the true numbers killed might be much lower, or much higher. It seems strange that in a country that loves both its cars and its wildlife so dearly, we have such a feeble understanding of the impacts of the one on the other.

Estimates of the numbers of reptiles and amphibians that we crush each year are even harder to come by, but the death toll must be enormous. As early as 1950, two researchers in Michigan recorded close to 1,000 dead anurans (frogs and toads) of eight species along a stretch of road less than a mile long. One study from Australia estimated that 10,000 frogs are killed each year per kilometre of road, although this figure was extrapolated from counts at a known hotspot for frog crossings. Another estimate from 1985, and thus surely an underestimate of today's body count, put the number of anurans killed on Australia's roads at 4.5 million per year, together with over a million reptiles. A short stretch of the BR-101 road in southern Brazil is thought to account for more than 15,000 reptile deaths each year. The scientific literature of roadkill is full of horrific dead-lists of reptiles and amphibians, and the total number killed on the world's roads each year must surely run into the billions.

As if the problem of accidental roadkill were not bad enough already, some motorists actually target animals on the road. Elegant experiments that compare vehicle strike rates of model snakes and turtles with those of dummy objects, such as a length of hosepipe or a disposable cup, show that a minority of drivers go out of their way, quite literally, to kill animals on the road. This is particularly the case for snakes, which seem to be the prime target of these roadkillers, although even harmless turtles are targeted by some drivers. Whether they do this through loathing, boredom or indifference is not known. But if it helps to restore your faith in human nature, one study in the USA found that just as many drivers stopped to help the fake snake or plastic turtle safely across the road as swerved to hit it.

When it comes to estimating the number of invertebrates killed on the roads each year, we can do little better than guess. Counts of road-killed butterflies at several sites in Illinois led researchers there to guesstimate that 20 million are killed each week in that one state alone, half a million of them migrating monarch butterflies, a species now in deep trouble. Another US study found that dragonflies are particularly susceptible to roadkill, with perhaps 35 killed per kilometre of road each day in the summer. On the Japanese island of Hokkaido, researchers recorded over 5,000 dead insects per kilometre of road, summed over 12 visits, with butterflies, beetles, flies, dragonflies, bugs, grasshoppers and hymenopterans (wasps, bees and ants) predominating. Crudely scaled up, this equates to over 50 billion insect deaths per year across Japan's road network, but counts of insects found on the road are likely to greatly underestimate the true number of casualties; most will be blown away by the wind or the slipstream of cars, removed by scavengers or plastered on windscreens. A study from Canada estimated the number of pollinating insects killed along a two-kilometre stretch of road in Ontario, and rather boldly extrapolated from this tiny sample to propose that the number killed across North America each year might be in the order of hundreds of billions – and this was just for three groups of insects. An unpublished study from the Netherlands, which asked motorists to count the number of dead insects splattered on their car number plates and log the distance they had travelled, suggested that in this one country alone over a trillion insects might be killed by traffic each year.

As for the numbers of slugs, snails, worms, spiders, beetles, woodlice or centipedes being crushed under our tyres each year – nobody has even tried to guess.

For all the hundreds of roadkill studies that have been published since Lillian and Dayton Stoner set out from Iowa City in the summer of 1924, we still know precious little about the global scale of the problem, other than that it is *big*. We do know something, however, about what happens if we reduce our driving. The lockdowns brought about by the COVID-19 pandemic in 2020 offered wildlife a rare reprieve from roadkill (human road casualties fell too). Across the USA alone, reduced traffic flow during lockdown might have spared, however temporarily, the lives of tens of millions of larger animals, and who knows how many billions of smaller ones. There was a nearly 60 per cent decline in the roadkill mortality of mountain lions in California during this brief period of traffic reduction. Across a number of European countries, roadkill fell by as much as 40 per cent. The numbers of hedgehogs killed on the roads in Poland fell by half, as did the numbers of marsupials killed on Australian highways and the number of amphibians killed on roads in Maine, USA. What was a tragedy for people proved to be a temporary respite for the planet's beleaguered wildlife.[13]

The impacts of roadkill

If estimating the numbers of animals killed by traffic is a very inexact science, then it is more difficult still to assess how important a threat to wildlife roadkill actually represents. The carnage we witness on an almost daily basis might suggest that roadkill is an important cause of death for many species. But there is of course a strong bias here – we are much more likely to see a road-killed badger than one that has died of disease or starvation deep underground in its sett. In fact, the little evidence we have suggests that roadkill is far from being the biggest wildlife killer when it comes to human activities. In terms of the numbers of birds killed each year in the USA, roadkill may come below predation by pet cats (which might account for a billion or more birds), collision with buildings and windows or flying into powerlines. It has been estimated, by a *very* back-of-the-envelope calculation (certainly not one that the scrupulous Dayton Stoner would have approved of) that perhaps only one in a thousand of all the world's birds is killed on the roads each year, although this still clearly sums up to a vast number.

Even if hitting billions of animals with cars each year is not the worst thing we do to the world's wildlife, it is easy to assume that any species that loses vast numbers of individuals on the roads each year must be seriously threatened by such slaughter. Researchers often fall into this trap: I have lost count of the number of scientific articles I have read that present a long list of dead animals and conclude from this that roadkill represents a serious threat to their species' survival. But if those animals would otherwise have been killed by other causes, or if the rate of reproduction is sufficient to make up the losses, then the impacts of roadkill can be absorbed. The vast tragedy of human roadkill does little to slow our own species' population growth.

Tasmania has been dubbed the roadkill capital of the world; some estimates suggest that more animals are killed per kilometre of road on the island than anywhere else in the world.[14] Tasmanians joke that the best place from which to view the island's rare and endemic wildlife is a glass-bottomed bus. Sometimes the impacts can be devastating; the upgrading of a road in Lake St Clair National Park led within two years to the complete extinction of eastern quolls and a halving of the local population of Tasmanian devils. But although the number of wombats killed on Tasmania's roads each year is enormous, the species' population has more than doubled over the last four decades. High levels of roadkill do not always lead to high rates of population decline.

Clara Grilo, a world expert in road ecology at the University of Lisbon in Portugal, is at the forefront of research into how roadkill might affect different species' populations in the long term. What she has found is that the species that lose the largest number of animals to roadkill are not necessarily those whose populations are most threatened by it. Her research shows that rather than looking simply at the numbers of animals logged on dead-lists, it is necessary to take into account each species' inherent vulnerability to those losses by considering factors such as how quickly it can reproduce and how long individual animals normally live.

A global assessment of the risks of roadkill to rare mammals, led by Grilo, identified four populations of particular concern: maned wolves and little spotted cats in Brazil, brown hyenas in South Africa and leopards in northern India are all at acute risk of roadkill-driven extinction. Roadkill might pose a significant risk to around 120 species of mammals (around 3 per cent of all the world's mammal species), including over 80 that are already listed as being at risk of extinction from other causes. In Europe, the hazel grouse and the russet ground squirrel emerge as the bird and mammal species, respectively, whose populations are most threatened by roadkill. This does not mean that these are the species that are killed on the roads in greatest numbers; rather, they are those whose populations can least afford losses to traffic. In contrast the house sparrow, which suffers huge losses on the roads and takes top spot on many dead-lists, is among the species found to be *least* threatened by roadkill. The animals we most frequently see dead on the roads are therefore not usually those whose populations are most threatened by roadkill. As so often, the biased evidence of our eyes can generate a very misleading view of reality.

Roadkill may become a more severe threat to a species' survival if it disproportionately affects a particular segment of the population. Take, for example, the European glow-worm, which despite its name is actually a species of snail-eating beetle. Male glow-worms, which can fly, are attracted at night by the glowing bioluminescence of the females, which lack wings. To occupy new habitats and avoid in-breeding, female glow-worm larvae need to spread out from where they were born, but being flightless they can only do so on foot (of which they have six). The great majority of glow-worms moving over the ground are therefore females. Unfortunately, these insects appear to be fatally attracted to roads, perhaps because the tarmac is warm and allows the larvae unimpeded travel. Huge numbers of female glow-worm larvae are killed on Europe's roads each year, perhaps contributing to the species' widespread decline. The fact that the victims are all female has a much greater impact on the species' populations than if the same slaughter were spread between the sexes, or even (because a single male can fertilise several females) if the same high level of mortality were borne by the males alone. Road traffic might not kill a high proportion of all glow-worms, but their populations might be particularly sensitive to the loss of these larval females.

A heavier roadkill of females seems to be a common pattern in invertebrates. The light refracted by dark asphalt can resemble the horizontally polarised light reflected by water, luring swarms of female mayflies to lay their eggs on roads with the loss of whole generations of these insects. For a brief period at the end of each summer, when the tides are particularly high, female land crabs in South Korea swarm from the forests to the sea to release their eggs. To reach the waves they need to cross busy roads, resulting in massive mortality that is borne almost entirely by egg-filled females. As a result of this high female mortality, the sex ratio of adult land crabs becomes massively skewed towards males, with severe consequences for the future health and viability of the overall population. The same pattern has been reported in freshwater turtles; near roads, where large numbers of females are killed as they search out ponds for egg-laying, populations are heavily skewed towards males. Among mammals, the pattern is generally reversed. From otters and red squirrels in Britain to koalas in Australia and giant anteaters in Patagonia, the general pattern is for males to suffer higher rates of road death than females, probably because they tend to wander more widely.

Roadkill certainly represents a serious problem to some species, particularly larger ones that produce few offspring each year. It is listed as a significant threat to the survival of more than 20 federally listed threatened species in the USA. The survival of the last viable lowland population of tapirs in the coastal forests of Brazil is jeopardised by the loss to roadkill of just one or two animals a year. In the same region, the entire world population of the golden lion tamarin, a stunning but highly threatened little marmoset with fiery orange fur, is at risk from the widening of a road that runs through the heart of its small home. Another South American mammal, the extraordinary besom-tailed giant anteater, faces extinction in areas near roads, which kill a fifth of all animals each year. Some of Australia's populations of the huge southern cassowary are close to being wiped out by roadkill. The magnificent Amur tiger is threatened by the few roads that run through its remote Siberian range.

Generally, however, the most frequent victims of roadkill are not threatened or endangered species; they tend to be those that are most abundant in the surrounding area. A recent study in the UK, based on thousands of reports sent in by the public, identified a few declining species, such as the barn owl and hedgehog, among the most frequently reported victims. It is entirely possible that their poor conservation status owes something to high rates of roadkill. But the list was largely made up of species whose populations are stable or even increasing (blackbird, buzzard, red fox, polecat, otter), species that many people might wish were not doing quite as well as they are (woodpigeon, brown rat, magpie) and species that arguably shouldn't be here at all (pheasant, grey squirrel, muntjac deer, rabbit, all introduced to Britain from elsewhere). It is certainly not a list that suggests that roadkill is a major problem for lots of rare and threatened species.

CHAPTER 4

Living with Roadkill

I n 1934, a debate was called in the Houses of Parliament to discuss the shocking number of people being killed in accidents on Britain's roads, some 7,000 each year (nearly four times today's fatality rate). The politicians decided by a majority that a speed limit of 30 mph should be imposed in residential areas, a limit that applies to this day. But not everyone in the House was happy with this decision. The haughty, car-loving Lieutenant-Colonel John Theodore Cuthbert Moore-Brabazon MP, later to be ennobled as 1st Baron Brabazon of Tara, argued vehemently against the need for such limit: casualties, he argued, would fall naturally as people got used to vehicles and learned to get out of their way. Rather insensitively, the ever-scowling Brabazon bragged about the havoc that he and his cronies had wrought in the early years of motoring:

> No doubt many of the old Members of the House will
> recollect the number of chickens we killed in the old
> days. We used to come back with the radiator stuffed
> with feathers. It was the same with dogs. Dogs get out
> of the way of motor cars nowadays and you never kill
> one. There is education even in the lower animals. These
> things will right themselves.[15]

Reading this, it is easy to understand the often violent response of people living in rural areas to early motorists. But the point is an interesting one: can animals adapt to the threat of roadkill? Brabazon's examples were not good ones: dogs can be tethered or trained to stay off roads, and keepers of chickens no doubt quickly learned to cage their birds to stop them from wandering out in front of the bloodthirsty baron's car. These are not cases of animals learning to stay safe, rather of people learning to keep them safe. But pet cats cannot be trained to stay off roads, yet the numbers we see dead on the tarmac today are very much lower than they were in the early days of motoring. Dayton and Lillian Stoner counted more than 30 dead cats during their 1935 trip through sparsely populated rural America, an average of one every 60 miles. We would be shocked to see such high rates of feline roadkill today, even in densely populated, heavily trafficated modern Britain. The cats have clearly learned a thing or two about road safety over the years.

Many animals refuse to cross busy roads at any time, so they have had no cause to evolve anti-roadkill defences. Others, such as bears, can sense that roads are dangerous places and try to cross them at night when there is less traffic around. Some have developed more sophisticated strategies to avoid roadkill, though few have road safety skills to match those of our closest relative, the chimpanzee. Researchers studying the response of chimps to a new asphalt road in Uganda's Kibale National Park found that almost all of them quickly adopted the good practice that is drilled into human children from an early age – look left and look right before crossing.

Some birds appear to be quite good at detecting approaching cars and taking evasive action. American crows, for example, have learned that cars tend to stay in the same lane as they zoom along and plan their escape accordingly. If a crow is feeding (usually on roadkill) in the same lane as an approaching car, it either flies off or hops over to the other lane well before the vehicle arrives.[16] If, on the other hand, it is feeding in the opposite lane, it seems to know it can safely stay where it is and watch the car go by. Another member of the resourceful crow family, the Florida scrub-jay, may also learn to avoid cars. Scrub-jays nesting near roads tend not to live as long, on average, as those nesting in forests, but it is the younger birds that get hit by traffic. As birds get older they seem to become more road-savvy, and from the age of three years and over they are no more likely to meet a sudden death as birds nesting far from roads.

Some birds appear to be unable to judge the speed of an approaching vehicle and base their flight escape response solely on the basis of how far away the vehicle is; this works perfectly well if the vehicle is travelling slowly but it does not give birds enough time to escape if the car is travelling fast. However, other species may be able to judge how fast cars are travelling, up to a point at least. Turkey vultures, which owing to their love of feeding on roadkill often find themselves contemplating the issue of road safety, seem to be able to judge the speed of an approaching vehicle reasonably well up to around 60 mph and plan their escape accordingly, but they are much less successful at dodging faster vehicles. Some interesting research undertaken in France has shown that on roads with higher speed limits, birds take off from the road further ahead of the approaching car than they do where speed limits are lower. The key finding of this study was that the speed that individual cars actually travelled made rather little difference – birds feeding on roads with higher speed limits would take off far in front of an approaching vehicle even if it was travelling slowly. They therefore seem to learn which roads are *generally* fast and which are *generally* slow, and base their evasive strategy on that. The problem with this strategy is that if they become habituated to roads with lower speed limits, and assume it is safe to take off at the last minute, they become very vulnerable to cars speeding over the limit. A car going at 40 mph in a 30 mph zone might therefore be much more dangerous to birds feeding on the road than a car travelling at 80 mph on a road with a 70 mph limit.

Some animals, therefore, have behavioural defences that reduce their risk of roadkill, but whether animals can evolve in such a way that their risk of sudden death actually falls over time remains less clear. It is quite easy to see how a species might fairly rapidly change its behaviour though a process of Darwinian natural (or rather *un*natural) selection. Think of a species whose population initially contains some individuals that are road-crossers and others that are road-avoiders. If the road-crossers fall victim to traffic before they can breed, while the road-avoiders stay safely at home and pass on their avoider genes to the next generation, then the entire population will quickly shift towards road avoidance and roadkill rates will fall.

Unfortunately, we don't have enough data on long-term patterns in roadkill rates to know whether this happens or not. Even if we did, we would not usually be able to tell whether a drop in roadkill rates means that a species is adapting to the danger or simply that its population is falling, leaving fewer animals to be killed by cars. However, one of the most intriguing studies ever published in the crowded field of roadkill research suggests that powerful evolutionary forces are indeed at work along the highways.

Working in Nebraska, husband and wife team Charles and Mary Brown undertook a long-term study of roadside colonies of a bird called the cliff swallow. As their name suggests, cliff swallows have an affinity for rocky crevasses and they have taken advantage of traffication by nesting under road bridges and in concrete culverts below the highway. The sudden appearance of miles of potential breeding habitat has opened up huge new areas of the country for them to breed in, and populations of cliff swallows have risen in many places as a result. On the downside, in taking advantage of this opportunity they are forced to live in very close proximity to traffic throughout the breeding season, and so they are exposed to a potentially high risk of roadkill. Remarkably, however, the Browns found that even though their population of road-nesting cliff swallows more than doubled over the 30 years of their study (and traffic volume presumably also increased greatly), the number of birds being killed on the road steadily fell.

Thankfully, the Browns had decided very early on in their study to collect and measure all the swallows they found dead on the roads. By comparing measurements of these roadkilled birds with those of living ones, they found that the swallows hit by cars tended to be those with longer than average wings. Furthermore, they found that the average wing-length of the whole roadside population became shorter over time. In birds, shorter and more rounded wings confer better manoeuvrability, particularly the ability to fly vertically upwards (this is why sparrowhawks, which rely on agility to catch their prey, have blunt, rounded wings, whereas falcons, which use speed, have long, pointed ones). The Browns' explanation for the patterns they observed was that shorter-winged cliff swallows are better able to get out of the way of speeding vehicles than the long-winged birds, and are therefore more likely to survive to pass their genes on to the next generation. The offspring of short-winged birds will themselves tend to be short-winged, so over time the population became increasingly dominated by short-winged birds better able to out-manoeuvre traffic, and roadkill rates fell as a result – an extraordinary illustration of rapid natural selection driven by roadkill.

So far as I am aware this study, published in 2013, is the only demonstration of rapid natural adaptation in response to the newly emerged threat of roadkill ever published, but there must be many more examples out there awaiting discovery. This, surely, will prove to be a fruitful field of research for aspiring young road ecologists.

On the other hand, there is also evidence that some animals might actually become *more* vulnerable to roadkill over time. Studies of the venomous copperhead snake in the USA have found that those living near busy highways secrete less of the stress hormone corticosterone into their bloodstream when they are faced with potential danger. The function of corticosterone is still poorly understood, but it probably plays an important role in increasing an animal's ability to recognise and respond to danger. It seems that life near busy roads has deadened the copperheads' ability to sense danger, either because they have become acclimatised to the roar of traffic or because the stress of roadside life has impaired their ability to produce corticosterone. Either way, their reduced ability to sense roads as a source of danger is likely to make copperheads more willing to cross and thereby end up as roadkill.

Just a wing

Every cloud, they say, has a silver lining. The blood-spattered tragedy of roadkill at least offers an opportunity to gather from the dead information that can be very hard to collect from the living. Animals that are hard to find, difficult to identify or dangerous to handle in life can yield a wealth of valuable scientific information when picked up limp and lifeless on the roadside.

Researchers working in the planet's most biodiverse region, the Tropical Andes, use roadkill as a way of learning more about the distribution and numbers of rare and threatened animals, even finding, flattened on the road, a species of snake entirely new to science. Another new species, this time a bird, was discovered in 1990 on a road in a remote park in southern Ethiopia. Even though only a single wing was left of the corpse, it proved to be so unlike anything else that the Nechisar nightjar was added to the list of the world's species without anyone knowing what the rest of the bird looks like. The nightjar's scientific name, *Caprimulgus solala*, reflects its damaged state – *solala* means 'just a wing'. Subsequent searches of the area have failed to find living birds, but it is possible the bird was struck far away and the wing was carried by the vehicle to where it was found. In 1996, an Indo-Chinese rat snake was found dead on a road in Borneo – the first record of that species on the island for nearly a century. More remarkable still was the discovery in May 2021 of another roadkilled snake, this time on a road in Assam, north-eastern India, that proved to be only the third ever record of a species called the blue-bellied kukri; the only two other specimens known to science had been collected 112 years previously, and nearly 300 km away. The first golden nightjar ever to be recorded north of the Sahara was struck by a birdwatcher's car on a road in Morocco in 2015. Closer to home, a pine marten found dead on a road in Wales in 2012 was the first record of that species in the country since the 1970s.

Victorian naturalists used the shotgun to collect vast numbers of interesting specimens for their natural history museums and private collections, material that still underpins much of what we know today about the natural world. Now, the car does that work for us.

The red-headed woodpecker, the most frequently recorded victim of roadkill of Dayton and Lillian Stoner's journey of discovery through Iowa in 1924, but strangely absent from their much longer trip in 1935, would become the first bird whose population decline would be charted through changes in roadkill rates. Perhaps more than anyone else, including even the Stoners themselves, the title of 'king of the dead-listers' belongs to A.W. Schorger, simply on the basis of the amount of driving he did. In an article published in 1954, he presented his roadkill counts gathered from nearly 50,000 miles of driving between 1932 and 1950. His observations showed a steady decline in the number of red-headed woodpeckers he found dead on the roads over this period, from nearly 50 a year in the early 1930s to just a handful in the late 1940s. These observations were backed by other evidence – there is no doubt that numbers of this stunning bird have declined greatly over time. But whether this was due to the effects of roadkill directly, or whether it merely reflected a decline in the species' population brought about by other causes, is unclear.

Now researchers all around the world are starting to glean information about trends in the populations of rare or hard-to-find animals by looking at changes over time in roadkill rates. This method might be particularly useful for tracking the arrival and spread of invasive species, which often follow roads and so are killed on them in large numbers. A black-headed python, a species endemic to northern Australia, was killed on a road in the Republic of Korea in 2019, alerting conservationists to the possibility that the species might have established a breeding foothold in the country through the release or escape of pet snakes. Invasive snakes can be absolutely devastating for local wildlife, and the sooner the problem is spotted the easier it is to deal with. In New Zealand, changes in the numbers of roadkilled possums, rabbits and hedgehogs give conservationists working to mitigate their impacts on native wildlife a chance to see how effective their actions are. The results are not always encouraging: rabbit haemorrhagic disease was deliberately introduced to New Zealand in 1997 to control the numbers of this cute but highly damaging invasive species, but roadkill rates showed that it had little impact; rabbit numbers measured by roadkill counts continued to increase even after the disease was introduced.

Scientists have realised that there is no better group of people to collect data on roadkill for them than drivers. Mobile technology now allows motorists to become monitors. In the UK, schemes such as Project Splatter and Mammals on Roads have already collected data on many thousands of victims. Examples of similar schemes elsewhere in the world include the Taiwan Roadkill Observation Network, the Australian Roadkill Reporting Project, Animals Under Wheels (Belgium), Srazenazver.cz (Czech Republic) and the grandfather of them all, the California Roadkill Observation System (whose online gallery of photos is a spatterfest of gore and entrails best avoided by those of a delicate disposition). A citizen science project that aims to gather information on Australia's wombats, whether dead on the road or alive next to it, has the wonderful name WomSAT.

Participants in these schemes are often provided with phone apps to report their sightings. They can then simply record each carcass they encounter, letting the technology note its precise location and time. Experts, or even Artificial Intelligence, can later confirm the identity of the victim from photos. The main aim of these schemes is to identify hotspots of roadkill so that measures can be taken to reduce the problem, but no doubt they also help to raise public awareness – as the old mantra of citizen science goes: if people start to count, they start to care.

Roadkill also yields ample material for studies on the genetics, ecology, diet and health of wildlife populations. Enterprising researchers in the Bahamas, for example, have collected large numbers of flattened Bahamian boas and racer snakes to study the internal parasites that they carry – how else could we ever know that female racer snakes are unwitting hosts to larger numbers of parasitic worms than males? Australian scientists studying a snake called the dugite have used roadkill to discover that the young snakes, which feed largely on small reptiles, have thick, blunt fangs that will not break when they hit tough, scaly skin, whereas the adults, which feed largely on mammals, have narrow, sharp fangs that are perfect for injecting venom into soft, warm flesh. Examining the finer details of the dentistry of these deadly snakes in living animals would be a very hazardous undertaking, but roadkill makes the task much easier. Road-killed deer carcasses are routinely tested for chronic wasting disease (CWD) in the USA: because deer with CWD are more likely to be struck by vehicles than healthy animals, roadkill offers a particularly effective way to map the spread of the disease. In Poland, researchers collect the toes of road-killed toads, not for making magic potions but to work out how old the animals are (by looking at bone growth rings) and thereby learn something about the animals' longevity. Professor Tim Birkhead, who has pioneered research into the extraordinary variation in the shape and size of avian sperm, extracted many of his samples from birds that he picked up off the tarmac. Remarkably, he found that the sperm he dissected from the testes of road-killed birds were still capable of swimming strongly even after the trauma of impact and several days of refrigeration.

The only limit to the use of roadkill in scientific research is, it seems, the imagination of the researchers.

Life and death in the fast lane

There is a human dimension to all this slaughter too: for every twisted animal corpse on the highway, there is a driver. We are both perpetrators and, to a certain extent, victims of this killing. In the USA, a country with an abundance of large mammals and fast highways, around 200 people are killed each year in collisions with animals. Some drivers, as we have already seen, intentionally target animals on the roads, but many others swerve or brake hard to avoid them, sometimes causing fatal accidents. Collisions between vehicles and wildlife cause damage to people and property estimated at billions of dollars each year. These human and economic costs partly explain the continued pre-eminence of roadkill in the field of road ecology.

Whether or not roadkill represents a major threat to the conservation of the planet's biodiversity, it is undoubtedly a hugely significant animal welfare issue. The Canadian author Timothy Findley has written movingly of encounters with flattened fauna during his travels:

> The dead by the road, or on it, testify to the presence
> of man. Their little gestures of pain—paws, wings
> and tails—are the saddest, the loneliest, most forlorn
> postures of the dead I can imagine. When we have
> stopped killing animals as though they were so much
> refuse, we will stop killing one another. But the
> highways show our indifference to death, so long as it is
> someone else's. It is an attitude of the human mind I do
> not grasp.

Yet our empathy for the sufferings of the billions of animals that die or are injured on the roads each year, and the fate of their dependent young, appears to have been reduced by our daily exposure to roadkill. In 1938, James Raymond Simmons made a perceptive observation: 'One reason why we have not encountered a widespread protest over the increasing loss of wildlife as a result of the modern car, the improved road and the lust for speed is this: the killing of a creature on the road is not a deliberate or malicious act.' We have seen that this is not *entirely* true, particularly in the case of snakes, but the fact that roadkill is at least largely unintentional must go some way towards explaining why, when there are so many protest groups targeting cruelty to animals held in captivity, there is none fighting for the accidentally roadkilled. Crowds of banner-waving protestors have closed roads in the name of many different environmental causes but never, somewhat paradoxically, to prevent roadkill. This might be because, in the words of anthropologist Jane Desmond:

> these animal lives have little value for most of the
> population … as these animals are unowned, lacking

in monetary or emotional value, not pets or livestock, and without the charismatic following that megafauna like elephants and lions in zoos receive. This calculus of devaluation clears the way for such carnage to be ignored in public discourse and legal venues, to be out of mind while insistently in sight.

But the first law of roadkill is that it is not an accident of random chance, not something we can simply put down to haphazard chaos beyond our control. Instead it is the product of known statistical probability, and as such it can be predicted and thereby reduced. There are plenty of things we can do to lessen the appalling toll of fear, suffering and death that our desire for convenient travel imposes on our fellow animals, as we shall see.

Roadkill is the most visible and the most poignant outcome of traffication. Cars kill billions of animals every year, and may threaten the very survival of a few rare and vulnerable species. Some local populations of sensitive animals such as amphibians and reptiles might be wiped out altogether by collisions with vehicles. But the species killed in largest numbers on the roads are generally common and capable of absorbing the losses.

Animals killed on the roads in large numbers all share a common characteristic – they cross roads frequently. And to cross roads frequently, they must live in close proximity to roads and be at least partly tolerant of the disturbance and pollution caused by road traffic. And if they are capable of surviving in heavily trafficated landscapes, then perhaps they are not the species that are being worst affected by our love of the car.

Before the 1980s, almost everything that was written on the ecological damage wrought by road traffic was about roadkill. But the dead-list mania sparked by the Stoners' first article in 1925 may have blinded scientists to the fact that roadkill is only part of the environmental toll exacted by cars. Even today, roadkill dominates the literature of road ecology. But the direction of research is slowly changing, because it is becoming increasingly clear that roadkill is only part of the problem posed to wildlife by traffication, and perhaps not even the most important part.

The slaughter of roadkill is monstrous, but it now appears that the animals at greatest risk from traffication may actually be those that never cross roads at all.

CHAPTER 5

Traffic Islands and Invasion Highways

Why did the chicken not cross the road? Why did the pheasant cross the road? Why didn't the lizard cross the road? Why did the snake cross the road? Why didn't the turtle cross the road? Why did the bear cross the road? Why did the chipmunk not cross the road? Why did the elephant cross the road? Why did the bat not cross the road?

This is not the setup for a string of weak jokes, although these questions clearly owe everything to a long history of chicken-related gags. In fact, they are all taken from the titles of serious scientific articles that analyse road-crossing behaviour in wild animals. In the first of them, the chicken in question is not the domesticated variety but the greater prairie-chicken, a wild and rare species restricted to the shrinking native grasslands of the USA. These titles may sound frivolous, and they certainly suggest that road ecologists need to find themselves some new cultural references, but the questions they pose are important. The decisions an animal makes when it arrives at the roadside might be the most important in its life, because they could be its last.

It is clear just from the way these questions are posed that animals differ in their road-crossing behaviour: four of them ask why the animal in question did cross the road, and five ask why it didn't. The subtitle of another scientific article, borrowed from a very different cultural heritage, sums up the issue well: 'To cross, or not to cross?' That is indeed the question, but it does not have a simple answer. There is more than one way of crossing a road – and more than one way of not crossing it.

We know this from our own behaviour. When we come to a narrow track or lane we walk across it as if it weren't there, trusting that its overgrown appearance and pitted surface are guarantee enough that we won't get flattened by a passing juggernaut. On reaching a more serious-looking road we might pause on the verge to check for danger, and then stop and check again when we reach the middle of the road. Or we might speed up and cross over quickly without stopping. If the road is really busy, and it looks as if no amount of pausing or speeding up would make the crossing safe, we stop and think again. Either we decide that we will never be able to cross it and look for other ways over – a bridge or an underpass perhaps – or we come back later when the traffic has died down a bit and run through all the various options again.

The careful research that sits behind the chicken-gag studies, and a host of other scientific publications with less clichéd titles, shows that animals adopt exactly the same range of strategies. When it comes to their road-crossing behaviour, we can identify four broad groups of animals.

First, there are the blind-crossers. These animals appear oblivious to the danger posed by the road and move across it as readily as they would through any other sort of habitat. Many of the animals falling into this category, mostly insects and other invertebrates, are simply unprepared by their evolutionary history for the threats posed by lumps of speeding metal. A moth that is fabulously well adapted by millions of years of natural selection to detect and evade the echolocating clicks of a hunting bat may have no defence against a rapidly approaching car. Blind-crossers include vertebrates too; some snakes appear to cross roads as if they were not there, and hunting barn owls allocate so much of their sophisticated sensory software to finding near-silent prey along noisy roadside verges that little is left over to detect and avoid traffic.

Some blind-crossers are driven across roads by internal forces beyond their control. In a few places each year, most spectacularly in Cuba, South Korea and remote Christmas Island, egg-filled land crabs swarm in their billions across roads, pulled towards the ocean by the swelling weight of tides and the even greater gravity of their evolutionary past. Unfortunately for them, road avoidance plays no part in this irresistible choreography, and uncountable numbers are crushed. The roadkill rate of blind-crossers is usually high, sometimes catastrophically so, and it increases steadily with the volume of traffic.

The second group are the pausers. Pausers will cross roads but, unlike the blind-crossers, they are not entirely oblivious to the threats facing them. If they feel threatened as they cross they deploy their anti-predator defence, which is to freeze. If they do so on the side of the road, and proceed only when the danger has passed, they may survive better than the blind-crossers, but freezing in the middle of the road is a suicidal strategy against cars. Many amphibians appear to be pausers; a Canadian study, for instance, found that more than 80 per cent of road-crossing frogs, toads and salamanders responded to an approaching car by stopping dead in their tracks (usually in both senses). Pausers also include animals that place a sadly misguided confidence in their defensive armoury, such as hedgehogs, porcupines, armadillos, tortoises, freshwater turtles and scorpions, and animals that play dead to deter predators, such as opossums and some snakes. Playing dead on a road quickly results in *being* dead on a road.

The third group are the speeders. Like pausers, speeders can sense the danger of traffic, or the unnaturalness of the road surface, but will nevertheless attempt to cross. The difference is that their response to approaching danger is not to freeze but to speed up and get across as quickly as possible. Studies that have followed the movements of a range of animals, from dragonflies to kangaroos to elephants, show that when they get to a road they put on a spurt to get across it quickly. This does not always save them, but it certainly helps; one of the main factors determining how likely an animal is to be killed on a road is how fast it crosses.

Finally, there are the road avoiders. These are animals that do not cross roads at all, or cross only the smallest and quietest of them. Their cautious behaviour means that, unlike the other three groups, road avoiders do not feature highly on roadkill dead-lists. Avoiders include some large mammals, many small mammals, some reptiles, certain bats and birds, most terrestrial molluscs and most ground-living insects. We don't know how most species respond when they come to roads, but further research may find that the avoiders make up the largest of the four groups.

Avoiders are not usually incapable of crossing roads; they simply prefer not to. Experiments carried out in Boston, Massachusetts, showed that when bumblebees feeding on flowering sweet pepperbush plants were caught and carried across a road, they soon flew back to their original territory. They were, therefore, perfectly capable of crossing the road. However, when the researchers removed the flowers from a pepperbush, forcing the bees to move on, they almost always switched to another bush on the same side of the road, even if there was a closer one on the other side. While they could cross the road, they appeared very reluctant to do so. Bumblebees, or at least these particular bumblebees, are avoiders.

It seems unlikely that a bumblebee can detect an oncoming car as a collision risk. It is probably something else about the road that deters it from crossing – its noise or petrochemical smell perhaps, or the unnatural radiation of heat or glare from the surface, or simply its unfamiliarity. Some animals avoid crossing roads because the open expanse of tarmac offers them no cover or boltholes to escape from predators; many small mammals appear to be avoiders for this reason. Sometimes, even small differences in the surface texture of roads can make a big difference. Stephens' kangaroo rat is a bizarre and rare little rodent, restricted to parts of southern California, that bounces along, like its far bigger (but biologically unrelated) Australian namesake, on outsized hind legs. Dirt roads present no obstacle to these animals, and may even help them to get around, but gravel and tarmac surfaces are strictly avoided and form barriers to the kangaroo rats' movements. Another staunch avoider, the dunes sagebrush lizard of the southern USA, will not cross even the narrowest of roads simply because it hates walking over anything that isn't its beloved sand. The road surface itself, therefore, often acts as a deterrent to avoiders, irrespective of the traffic it carries.

It might not always be easy to guess which of these four strategies a particular species will adopt, because even closely related species can behave very differently. Barbastelle bats, which forage on flying insects, readily cross motorways when feeding or moving between roost sites – they are blind-crossers, but usually safe from vehicles in the high airspace they inhabit – whereas Bechstein's bats, which feed by plucking insects from vegetation, are strict road avoiders. Beetles of dry, open habitats are blind-crossers or speeders, whereas forest beetles, particularly flightless species, are no-nonsense avoiders. Smaller snakes tend to be avoiders and larger ones tend to be speeders, and among the latter, non-venomous snakes speed across the road faster than venomous ones.

Many animals seem to be able to judge how wide a road is, or how busy, and adapt their behaviour accordingly. In what must have been some of the most tedious ecological fieldwork ever undertaken, researchers working in central Sweden spent four months following the movements along roadside verges of a little mollusc called the copse snail. The most energetic of these snails covered marathon distances of up to 14 m during the study, more than enough to take them across the adjacent thorough-fare and back again, but not one of them would cross even so much as a dirt track. A narrow footpath, however, proved no obstacle to them. Copse snails are avoiders when it comes to tracks and roads, but blind-crossers (or perhaps even speeders, in their own minds at least) when it comes to footpaths. Hedgehogs, badgers, wild boars and many other mammals are speeders when they come to quiet roads and avoiders when they meet busy ones.

An animal's decision will also be based upon circumstance, motivation and experience. If it is being chased towards a road by a predator it may become a temporary blind-crosser, perhaps seeing the danger ahead as a last opportunity to throw off its pursuer. A starving animal is more likely to accept the risk of a motorway crossing than a sated one, and an animal that knows from past experience that rich resources lie on the other side may take its chances on the road more readily than a naïve one.

The picture that emerges is that animals behave very differently when it comes to crossing quieter roads – some will cross, in their own chosen ways, and others such as the copse snail will not – but as the width of the tarmac and the volume of the traffic increase, so more and more of them become avoiders. The very busiest highways will deter almost all attempts at crossing, by any species: even the most enthusiastic blind-crossers will usually think again when faced with a hostile barricade of eight lanes of tarmac, blinding light, roaring noise and petrochemical stench. For most animals, busy roads form impenetrable walls across the countryside.

This might seem like a good thing, because if no animals attempt to cross these violent highways then none will be flattened. The avoiders, you might think, should fare better in our heavily trafficated land than the blind-crossers, the pausers or the speeders, because their roadkill risk will always be lower. Unfortunately, things are not quite as simple as that, because it turns out that *not* crossing roads brings with it a range of quite different environmental problems. Indeed, despite being relatively safe from the threat of roadkill, the avoiders may prove to be the biggest losers of all in our increasingly trafficated world.

The warning of Barro Colorado

We no longer live in an intact landscape: a century of traffication has put paid to that. Take another look at a map of Britain's roads (Map 1, p. 36) and this time try to see it in negative, not as a network of roads but as a patchwork of islands surrounded by roads. The world has forty or so landlocked countries, those with no sea border. Two of them, as any experienced pub-quizzer can tell you, are double landlocked: Liechtenstein and Uzbekistan are entirely surrounded by other landlocked countries.[17] But we are all roadlocked – ten, fifty, perhaps a hundred times over. This has profound implications for our society: roads form physical barriers to the lateral movement of people, money and ideas; they subdivide cities and compartmentalise communities like a web of Berlin Walls. Sometimes this community severance, as it is known, has been done deliberately; in the USA, major highways have been routed intentionally to form a barricade between communities of different ethnic origin or socioeconomic status. So serious is the problem that in March 2021, President Joe Biden proposed spending $20 billion to help reconnect neighbourhoods that have been riven by highways. The question of whether roads do a better job of connecting people or dividing them is still an open one.

Our wildlife, too, has to find ways to survive in this mosaic of road-partitioned fragments. Only recently have we started to recognise just how damaging this cookie-cutting of our countryside could prove for wildlife, because its effects may take years or perhaps decades to become apparent. To understand what happens over such time scales to animals and plants living in landscapes that have been chopped into little pieces, we need to visit the tropical forests of Central America.

In 1913, engineers working on the construction of the Panama Canal dammed the Chagres River, leading to the flooding of a huge area of Panama's lowland forests and the creation of Gatún Lake. This was the world's largest artificial waterbody until Lake Mead was created by the construction of the Hoover Dam in 1936. Today, a ship's passage through the Panama Canal, between Colón on the Caribbean coast and Panama City on the Pacific, requires 20 miles of navigation across this great expanse of water.

Not all of the former valley of the Chagres River was inundated, however, because the highest hilltops remained poking out above the floodwater and became isolated as islands in the new lake. The largest of these, at 15 sq. km, is Barro Colorado, which in 1924 (the same year that Dayton and Lillian Stoner set out on their historic roadkill-counting drive across rural Iowa) became home to what is now one of the oldest and most famous biological research stations on the planet. The Smithsonian Tropical Research Institute, which runs the station, provides facilities for hundreds of visiting scientists each year. There is barely an animal or plant on the island, it seems, that is not the subject of someone's research.

This wonderful wealth of biological information, which stands in stark contrast to our woeful ignorance of most of the rest of the planet's tropical biodiversity, offers a rare opportunity to see what happens when parts of what was once a continuous, interconnected landscape suddenly become isolated as fragments. And what happens, it rather depressingly transpires, is that the newly isolated populations of many of the species stranded on these fragments enter terminal decline and eventually, over years or decades, die out altogether. Barro Colorado has been the subject of intensive conservation action that has led to a doubling of the area of mature rainforest on the island and a complete ban on hunting. Yet over a quarter of the forest bird species present when the rising floodwaters first turned it into an island have since been lost.

Barro Colorado has become a textbook illustration of the perils of habitat fragmentation – the breaking up of continuous landscapes into smaller, isolated pieces – but what has happened there is far from unique. Studies of fragmentation from around the world have shown exactly the same pattern, and not just for birds: mammals, insects, amphibians and other groups of animals are all affected in exactly the same way. The local numbers of animals caught in the fragment might not, at first, be any lower than they were before they were surrounded, but isolating populations in this way proves to be a mortal wound that leads to extinction decades later.

The process that drives this slow death was identified in the 1960s by two brilliant American ecologists, Robert MacArthur and E.O. Wilson, whose book *The Theory of Island Biogeography* (1967) must rank as one of the most important ever published in the natural sciences. The heart of the problem is that fragmentation splits a large, freely intermixing population of a species into numerous small isolated ones, each of which is more likely to become extinct than the original. A fragmentary population of just a few animals or plants is much more likely to die out through chance events, such as a series of severe droughts or the passage of a hurricane, than a population that numbers hundreds or thousands spread over a much larger area. It might take many years for the forces of extinction to fall into perfect alignment and wipe out the last few individuals of a particular species in a fragment, but the fate of the stranded population was sealed the moment the waters first closed around it.

There is another important feature of fragmentation, which is that once a species has been lost it generally does not return: the loss of species from fragments is a ratchet of extinction, turning only one way. This is because the species that tend to be lost are the avoiders – they rarely, if ever, cross whatever barrier it is that encloses the fragment. At its narrowest, the stretch of water that separates Barro Colorado from the adjacent mainland is only 250 m across. Yet the species that have been lost from Barro Colorado's forests have been unable to cross this narrow stretch of water and recolonise the island, even though many remain common on the other side. These species are dyed-in-the-wool avoiders when it comes to crossing water, or indeed anything else that is not forest. This seems to be a general characteristic of shy forest animals – even an old, overgrown logging road might represent an impenetrable barrier to them. It is a strategy that clearly works for them when conditions are good, because many tropical species adopt this behaviour, but it entails the irreversible loss of species in fragments such as Barro Colorado. Over time, each fragment slowly bleeds biodiversity but receives no transfusions, and the number of species it supports dwindles.

Another general law of fragmentation is that the smaller a fragment becomes, the more species that were originally present will be lost. If the floodwaters of Gatún Lake had risen higher and Barro Colorado been smaller, even more of its original complement of species would by now have died out. The smallest fragments may, in time, lose all of the species that once occurred there.

The lessons learned from Barro Colorado move uncomfortably close to home if we look again at Britain's road map (Map 1, p. 36). Try to picture these thousands of traffic islands as an archipelago of little Barro Colorados, each of them separated from its neighbours by the rising floodwaters of tarmac and traffic, and each slowly haemorrhaging wildlife. This is not a fanciful or alarmist comparison. Our biggest roads are not much narrower than the 250 m of water that separate Barro Colorado from the nearest mainland (indeed some mega-highways, such as the 26-lane Katy Highway in Texas, or the Monumental Axis in Brazil, are actually wider). And even a narrow road could make just as effective a barrier to wildlife as a much wider stretch of undisturbed water in a peaceful nature reserve. Furthermore, the smaller the fragment, the more impoverished its wildlife and the greater the rate of biodiversity loss – and almost all of our traffic islands are smaller than Barro Colorado. The problem is a global one: by one estimate, major roads alone have carved the planet's land surface into more than 600,000 tarmac-edged traffic islands, most of them further subdivided by smaller roads.

Roads are not, of course, the only obstacles to animal movement in our world. Rivers, mountain ranges and seas have divided populations of animals and plants for so long that those on either side have evolved into different species (a process known as vicariance). But nothing slices up and fragments the countryside into such tiny parcels of land as our road network, and nothing has done it so quickly. If the division of our landscape into thousands of little fragments has anything like the long-term impacts that isolation has had on Barro Colorado's once-rich biodiversity, the populations of many of our wild species might already be condemned, invisibly but irretrievably, to a long, slow slide towards extinction.

Living in fragments

The patient researchers who followed the slow-moving Swedish copse snails concluded from their research that even the quietest road cleaves apart populations on either side as effectively as would a river of salt. Assuming that British copse snails respond to our busier and denser road system in the same hardcore-avoider way as their Swedish cousins, then it is clear that we do not have one large, interconnected population of this pretty little mollusc but thousands of small, isolated ones. It would be a mistake to see a map of Britain's road network as a perfect illustration of this fragmentation because some roads will be permeable even to the sluggish movements of copse snails – through underground culverts, perhaps, or over bridges. Some snails will get moved between fragments by accident (perhaps, somewhat ironically, by crawling onto or into cars and being transported many miles). But the slimy wanderings of these gastropods must now be greatly restricted compared with the days before traffication. Once they have oozed their way slowly to the hostile edge of their traffic island, the snails now have nowhere to go but back again. All the resources that lie on the other side of those few metres of tarmac – food, mates, breeding sites, damp areas to shelter in during droughts, higher ground to escape frost or flooding – are lost to them forever. This may doom the trapped population to eventual extinction and the traffic island will lose its snails, perhaps irretrievably. Being an avoider can incur heavy costs from traffication, even if roadkill rates are low.

For larger animals, whose road-crossing behaviour depends at least partly on the number of vehicles using the road, the rising floodwaters of traffication continually reduce the size of traffic islands. Fifty years ago, their islands might have been defined by motorways, since only the busiest roads would have carried a volume of vehicles sufficient to deter all attempts at crossing. Motorways are few and far between in the landscape, so the fragments they enclose are generally large. But as other classes of road have become busier, turning more and more of them into impassable barriers, so the islands have become smaller and smaller. After a time, the risks of crossing a road to reach an island start to outweigh the benefits and the fragment is abandoned by the animals that once occurred there. Populations of larger animals in these heavily trafficated landscapes start to flicker and die out. Well-intentioned efforts to help these threatened populations by installing roadside fencing to reduce roadkill only serve to make highways even more impermeable.

Larger carnivores are particularly vulnerable to fragmentation by roads because they need huge territories. In the USA, there is already evidence that bears and large cats have become extinct within some traffic islands owing to barrier effects. The USA's population of grizzly bears is now limited to a few small, isolated groups of animals, mostly along the border with Canada. Their continued survival depends on the exchange of animals between these lingering pockets in the USA and the much larger Canadian grizzly population, both to maintain genetic health and to restock numbers. Unfortunately, this vital bridge has been severed right across southern Canada by a major highway. Another fearsome Canadian predator, the wolverine, stays well away even from roads carrying as few as 30 cars a day. Roads have also been identified as obstacles to the movements of tigers in India. As with the USA's grizzlies, the survival of small and isolated groups of this magnificent predator relies on the movement of animals between them, and the barrier effects of major roads could lead to their extinction. One of Europe's most endangered mammals, the Iberian lynx, may be extinct within a century for the same reason.

Even the largest of animals, living in some of the most untrafficated parts of the world, are constrained by roads. Studies that fitted tracking devices to African forest elephants in the Congo Basin found that roads outside protected areas represent almost impassable obstacles to wandering animals, probably owing to the use of these roads by poachers. The authors of this study concluded that uninhibited ranging by forest elephants might now be extinct as a natural phenomenon, and that if road-building schemes in the region continue along their current trajectory, forest wildernesses and the elephant populations they support will collapse.

Fragmentation by roads may affect not only populations of individual species, but also entire wildlife communities. Another study from Sweden (the Swedes are great road ecologists) compared the communities of wasps and bees on opposite sides of a busy highway. Sample sizes were small and only a single road was included in the study, but what the researchers found was deeply worrying. The total number of bees and wasps was no different on either side of the road, but the species present on each side were quite different. Some species occurred on both sides, but some, particularly the smaller species, were found only on one side or the other. The barrier effects of the road appear to have led to the creation of two quite different communities of insects, each of them less than the natural sum of their parts. This means that species that would normally occur together in the same natural community, with all its incredibly complex ecological inter-actions, are now parted from one another. Bees and wasps are important pollinators of some of our crops and play a significant role in pest control, adding to the potential problems. If this pattern holds generally (and more studies like this are surely needed), then our insect communities might have been fundamentally altered wherever there are roads. Which in Britain, as you can see if you look again at Map 1, is pretty much everywhere.

The flow of genes

Tracking the movements of animals is difficult – not many people have the patience, time and skill to follow snails or mice or wolverines or bears for long periods to see how often they cross roads. Sophisticated tracking devices can now be fitted to all but the smallest animals, but to do so you need first to catch the animal, and their cost limits the number of individuals that can be followed.

Thankfully for researchers wanting to study the barrier effects of roads, new methods have become available in the last few decades that allow the intermixing of populations of animals to be examined without having to follow their movements at all. By analysing the genetic makeup of different groups of animals, it is possible to see how much they intermingle. If the genetic signature of animals on one side of a road is pretty much the same as those of the same species on the other side, then they clearly interbreed regularly; the road is not an impermeable obstacle to movement or mating, even if many animals die in the attempt. But if the road stops movements between them, the rate of interbreeding falls and the populations on either side of the road start to drift apart genetically. The difference in the genetic makeups of two populations of a species on either side of a road can therefore allow researchers to estimate how often animals from one side of the road cross over and mate with those on the other, and thereby to work out how much of a barrier the road poses to them. Weeks of arduous fieldwork can now be avoided by taking a single drop of blood or a tiny sliver of skin from a trapped animal before releasing it back into the wild. Scientists can now isolate DNA from an animal's droppings, or from shed hairs or feathers, meaning that animals no longer even need to be caught for their genetic code to be read. And of course roadkill provides a well-stocked larder of genetic material for researchers to plunder.

Worryingly, the results of almost all these studies show that roads make very effective barriers indeed.[18] The copse snail, it seems, is far from unique in refusing to cross tarmac. Take the wood mouse, one of our commonest and most attractive wild rodents. Research undertaken in Spain and Portugal (this species occurs there too) found that wood mice living on one side of some fairly quiet roads are genetically more different from those on the other side than they are from wood mice on the same side of the road but over a kilometre away. In other words, a few metres of tarmac separate wood mice from each other more effectively than a kilometre or more of other habitats. The researchers also found that the level of genetic divergence between populations separated by a busy road was no greater than between those separated by quieter roads, showing that it is the physical structure of the road itself that stops animals from crossing, not the number of vehicles it carries. Tellingly, mice on either side of a road were genetically more similar to each other where culverts ran under the highway, providing them with an opportunity for crossing safely. The mice, and their genes, flowed freely under the road surface, but not over it.

Very similar results have emerged from a host of studies of other rodents, as well as snakes, salamanders, freshwater turtles, frogs, newts, bees, beetles, spiders, pollinating insects and many other types of animals. Researchers have been able to confirm that for many small species, as with the wood mice, it is the road itself that forms the obstacle, not the traffic it carries.

The genetic makeup of larger species is similarly affected by road fragmentation. Bobcats, coyotes and bighorn sheep in North America, koalas in Australia and red deer, roe deer and wild boar in Europe all reveal the same pattern of genetic divergence between animals on either side of busy highways. For these larger animals, traffic volume plays an important role in determining crossing behaviour: they are more prepared than smaller animals to cross roads if the volume of traffic is low, but very busy roads divide populations of large mammals just as effectively as they do those of tarmac-shy snails or mice.

If a road slices a population of animals into two, and slows or stops the flow of genes between them, two things will happen. First, through a process known as genetic drift, the animals on one side of the road become more and more dissimilar from those on the other side – the genetic makeup of the two newly separated populations starts to drift apart. This can happen surprisingly quickly. Brave researchers in the USA studying timber rattlesnakes, a species with the wonderful scientific name *Crotalus horridus*, found that within just a few years of the completion of a new highway there was significant genetic divergence between the rattlers living on either side. These snakes rarely if ever cross paved roads, and in heavily trafficated areas almost every snake's home range is now entirely bounded by roads; each is now a prisoner in its own traffic island.

Second, newly separated populations suffer a loss of genetic diversity; the smaller the population, and the longer it has been isolated, the greater the loss of genetic diversity. Populations of bighorn sheep can lose up to 15 per cent of their genetic diversity within just four decades of being isolated by roads. For animals imprisoned in traffic islands, this loss of genetic diversity further increases the likelihood of extinction, because it can reduce both their average lifespan and their capacity to reproduce. It also makes them less adaptable to new threats: a population with low genetic diversity will contain fewer individuals that are resilient to new diseases, or increased levels of fire or drought.

The barrier effects of roads can have serious impacts on plants too. Small rodents, which usually avoid crossing roads, are important seed

Part of Highway 401, Ontario, Canada – a significant obstacle to animals with wings and an impenetrable barrier to those without. As well as the wide expanse of tarmac and many lanes of heavy traffic, an animal crossing here would also have to negotiate several solid crash barriers, which add to the road's impermeability. Roads much narrower than this, even single-lane tracks, can effectively isolate populations of animals living on either side.
(Wikimedia Commons)

dispersers; research from China has shown that wild apricot seeds are moved along roadsides by mice and other rodents, but they are very rarely moved across even quiet country lanes. Important pollinators of plants, such as honeybees, bumblebees and hoverflies, are also resolute road-avoiders, meaning that the flow of pollen across roads is similarly cut off. Over time, the insect-pollinated plants living on either side of a road will also start to drift apart genetically, and floral communities will become more impoverished.

It is slightly disturbing to think that if we could analyse the genetic makeup of a big enough sample of wood mice or copse snails from right across Britain and plot the results on a map, we might see in it a pattern resembling our road network. Our love affair with the car is deeply encoded in our wildlife's DNA.

Invasion highways

There is a second part to this story, which crosses the first at right angles. Roads form very effective barriers to animals trying to move *across* them, but they can also make extremely efficient conduits for those moving *along* them. Animals that are able to follow highways can spread as far and wide as the road network itself, so long as conditions remain suitable for them. Unfortunately, the species that take advantage of these unimpeded channels of movement tend to be very unwelcome visitors.

Where they occur naturally, plant and animal species have had hundreds of thousands of generations, perhaps millions, in which to occupy all the habitat types and climatic conditions that they are capable of reaching and inhabiting; they have filled all the niches available to them. Their distributions wax and wane slowly over time, largely in response to fluctuations in climate, but at any given time a species has generally occupied pretty much everywhere that is both suitable and accessible.

The situation is quite different for species that are introduced by people, deliberately or accidentally, into wholly new parts of the world. For these non-native (or invasive) species, the scramble is on to invade all the suitable habitats they can reach: it is a biological land-grab. Unlike the shy forest undergrowth birds of Barro Colorado, unwilling to cross even so much as a dappled glade let alone an open highway, successful invaders are adaptable risk-takers, able to make use of any opportunities on offer. Roads can be very helpful to them when it comes to conquering virgin territory as quickly as possible. After all, reaching places quickly is exactly what roads were designed to do.

Toad on the road. The cane toad's invasion of Australia has been speeded by its ability to use roads, which have allowed it to spread westwards at 50 km a year to the huge detriment of the local wildlife, not to mention people's pets. Even passionate naturalists struggle to find much to love about the cane toad. *(Wikimedia Commons)*

The cane toad is a rather grumpy-looking, dinner-plate-sized amphibian that occurs naturally in the grasslands and forests of Central and South America, where it leads a relatively blameless, if somewhat gluttonous, existence. Its voracious appetite and willingness to eat almost anything it can cram inside its capacious mouth were noted admiringly by scientists looking for a way to control outbreaks of beetle pests in sugar cane plantations in far-off Australia. Rather than using chemical sprays, the scientists thought, let's see if this insatiable amphibian can solve the problem for us. After seemingly successful trials in a number of places, including Hawaii and the Philippines, cane toads were released into north-eastern Australia's sugar cane heartlands in 1935.

It proved to be a catastrophic mistake. The toads failed to control the pests they were put there to devour, but they proved to be very adept at guzzling up the native wildlife. Furthermore, cane toads are extremely poisonous, their backs covered with toxin-filled pustules; even their tadpoles are poisonous. Dogs are particularly susceptible to cane toad toxin and many pets have died after attacking these unfamiliar newcomers. Native predators too have been affected. All the species of crocodile and freshwater turtle that occur in Australia are capable of swallowing a cane toad big enough to kill them; monitors, goannas and other lizards are equally vulnerable. Among mammals, the northern quoll, a highly endangered species that looks like a cross between a cat and a spotted rat, appears to be at greatest risk. Cane toads are also very partial to dung beetles, hoovering up over a hundred of these useful and hard-working insects in a single sitting.[19]

Since its introduction to Queensland's tropical north-east, the cane toad has spread rapidly across northern and eastern Australia. Its southward march, which has taken it to Sydney and beyond, appears to be slowing down, but its range is still expanding westwards across northern Australia at 50 km or more a year.

This rapid rate of spread has been possible because of two adaptations. First, cane toads at the leading edge of the invasion wave have longer legs and are capable of moving faster than their more sedentary forebears behind them. Second, cane toads use roads. They move much more quickly along highways than they can through dense vegetation, and this is their preferred means of overrunning new areas. For an amphibian, the cane toad is unusually well adapted to life on the open road. Its poisonous skin means it does not have to hide from predators, and it is capable of surviving the loss of over half the water in its body (indeed, it is now invading even desert areas). Its only problem is the light traffic that passes along these remote highways. Thousands of cane toads are killed on the roads each year, many of them mown down deliberately by grieving dog-lovers, but the benefits of rapid spread along straight, unimpeded highways clearly outweigh the costs of roadkill.[20] Cane toads are not the only unwelcome and unnatural additions to Australia's fauna that use roads – dingoes, foxes and feral cats are all strongly attracted to quiet, unsurfaced roads, using them as convenient routes of access through dense woodland or scrub.

Invasive plants are also particularly good at using roads to further their spread. Roadside verges support few native plant species to compete with, few grazing animals and few insect pests, so the non-native plants that are able to survive here often do very well. A study undertaken in the remote and beautiful Kashmir Valley in the western Himalayas, an area that cries out for the use of the word pristine, found that 70 per cent of the plant species recorded along roadsides were not native to the region. A similar situation has been documented in other remote areas, such as the high Andes of Bolivia and Glacier National Park in Montana. Roads act as pipelines for the spread of non-native species through otherwise impassable mountain ranges, opening up huge new regions for invasion on the other side.

Like the cane toads in Australia, invasive plants have escaped their natural enemies and are free to keep spreading along roadsides as long as conditions remain suitable. But unlike the cane toads, they actually benefit from traffic – indeed, the heavier and faster the traffic, the better. Common ragweed (not to be confused with common ragwort) is a plant native to North America that can make life a misery for people allergic to its pollen (symptoms range from mild hay fever to severe asthmatic attacks). Ragweed has become accidentally established in many parts of southern Europe and is now spreading steadily northwards, helped by a warming climate. This northward spread is being speeded by road traffic. Under normal conditions, ragweed seeds tend to fall close to the parent plant and so it spreads slowly, just a few metres each year. On roadside verges, however, the air currents whipped up by speeding vehicles are capable of picking up seeds that have fallen on the tarmac and depositing them much further along the verge. This speeds up the rate of spread by a factor of ten or more, but only in the direction in which the traffic is moving. The faster the traffic, the further along the road the seeds are wafted.

More important still for the spread of invasive plants along roads is their ability to hitchhike long distances. Seeds have many adaptations that allow them to spread, or rather to be spread by others, far and wide. Many plants have seeds that are equipped with spines, hairs or hooks so they can attach themselves to passing animals. They can also attach themselves to cars and their passengers. Seeds can accumulate in many different places on a car, including under the chassis, in the tread of tyres, in radiator grilles, in wheel arches and on interior mats and upholstery. Any mud sticking to the outside of the vehicle is sure to harbour seeds; experiments have shown that seeds can be transported for hundreds of miles in this way. Over 600 species of plant are known to have car-dispersed seeds, many of them invasive and more than 100 classified as internationally important environmental problems. Indeed, an invasive species' success in infesting new areas depends in large part on how well its seeds are dispersed along roads. Most of the invasive plants that spread along roads are species that were first imported by the horticulture trade – and the fact that most cars begin their journeys from their drivers' gardens may be no coincidence.

The long-distance dispersal by road traffic of the seeds of invasive plants is not a rare, one-off phenomenon. An ingenious study undertaken in Berlin looked at the number of seeds found in a long stretch of subterranean motorway. Being in a tunnel, the seeds lying on the road there could only have arrived on vehicles, so the researchers could eliminate other sources. They estimated that up to 1,500 seeds are deposited on every square metre of road by vehicles each year, half of them the seeds of invasive species (including many known to cause problems for native biodiversity). Traffic does not just shed the occasional harmful seed; it produces a constant rain of invasive propagules along all our highways. Many of these will end up on the verge, take root and flower, and scatter a new generation of seeds ready to be picked up by passing vehicles and deposited miles further down the highway.

Other undesirables follow roads, too. Native earthworms were wiped out from most of Canada and the northern USA during the ice ages, but settlement of the continent by people from Europe brought with it non-native earthworms. These have invaded Canada's boreal forests along highways, entirely changing their ecology. The red imported fire ant (very unaffectionately known as the RIFA) is an aggressive little insect that packs a huge toxic punch, and in large numbers it can overpower and butcher animals very much larger than itself. Introduced into the port of Mobile, Alabama, in the 1930s, it has spread along roadsides across much of the south-eastern USA and recently invaded the Florida Keys, threatening many rare species there. Mildews, moulds, root rots and other plant diseases also spread along highways. Of more direct concern to ourselves, so too do diseases of humans: studies from Africa, for example, have shown that roads speed the spread of rabies-carrying dogs.

The first cut

There is some debate about whether or not we should count ourselves as an invasive species, but whatever the niceties of the philosophical arguments there can be no doubt that of all the world's large terrestrial species, we (and our ragbag of camp followers – mice, rats and the like) are the most widespread. From our ancestral homeland in north-eastern Africa we have spread out to occupy most parts of the globe, and our impacts have been profound. Some scientists have declared that the Holocene, the geological epoch that has persisted for 12,000 years since the last major ice ages, is over, and that we have now entered a new epoch, the Anthropocene – the age of mankind. In 2019, the Anthropocene Working Group, working under the auspices of the seductively named Subcommission on Quaternary Stratig-raphy, voted in favour of recognising the Anthropocene as a new and valid epoch, with a nominal start year of 1950. Those in favour of recognising the Anthropocene argue that human activities have reached such a scale that they are significantly impacting the planet's geology. The geologists of the far future will discover an abrupt transition in the strata they study, finding clear evidence after 1950 of profound changes in global temperatures, chemistry and habitats. Their palaeontologist colleagues will find evidence of a massive decline in biodiversity and a spike in extinction rates. Something else these far-future excavators of our world will find is roads; lots and lots of roads. Like the thin layer of dark, iridium-rich ash that records the meteor impact that killed off the dinosaurs, the boundary of the Holocene and the Anthropocene will be marked in the future geological record by a narrow seam of tarmac.

In his book *The Car*, the British historian and journalist Bryan Appleyard describes the vehicle as 'the Anthropocene's battering ram'. It is a clever metaphor: if unspoiled nature is a citadel, then the car is the siege engine that has breached its walls and allowed the invading hordes access to its inner sanctums. Where roads lead, people, pests, development and agriculture usually follow. So synonymous are roads with a wide range of other threats to wildlife that wilderness areas, those with very few human impacts of any sort, are often referred to as *roadless areas*. This is more than simply a convenient term of reference; it has historical origins, since the early wilderness movement in 1930s USA, which pushed for the establishment of that country's great national parks, was stimulated in large part by concerns about the rising dominance of the car.

Even though they occupy only a tiny proportion of the landscape themselves, roads can start a contagion of habitat destruction that spreads over huge areas. It is a pattern that geographers have documented a thousand times, one that satellites record on a daily basis. First, a new road is cut through a pristine area to link two distant towns. Then the narrow strip of road becomes a broad band of development as the adjacent land is opened up to exploitation. Hunters, poachers, miners, loggers, farmers, builders, colonists, speculators and land-grabbers gain access to hitherto untapped resources. Fires, pollution and invasive species spread outwards with them. Eventually, the distance to the initial road becomes too far to sustain further commercial exploitation, so new roads are built perpendicular to the first, and fresh bands of development begin to spread outwards from them. The process continues, new roads spreading like fungal mycelia, until all the undeveloped land for miles around has been consumed. A huge swathe of natural habitats may eventually be damaged or erased entirely by the building of a single road; as the old song goes, the first cut is the deepest.

In small, heavily trafficated countries such as the UK this damage was wrought long ago; the only roadless areas of any size now are those whose steep slopes render them inaccessible to traffic, or whose soils or climate are unsuited for commercial use. Our country has long been saturated with roads; most of the new highways built today run close to existing ones, usually in the vain hope of relieving them of some of their congestion. But in other parts of the world, extensive roadless landscapes still survive. Only around 20 per cent of Britain's land area lies further than 1 km from a road (see Map 3, p. 38), and across Europe as a whole around 40 per cent of land is similarly road-free. But across the planet's entire land surface, with its extensive deserts, mountain ranges, boreal forests and permafrost, around 80 per cent of all land still lies further than 1 km from the nearest major road.

The biggest of these remaining roadless areas are found, perhaps unsurprisingly, in the vast boreal forests and arctic barrens of northern Canada, Greenland and Russia, and in the great deserts and arid zones of the Sahara and Arabia, central Asia, Mongolia and Australia. These generally infertile and climatically hostile areas are unlikely to become heavily trafficated in our lifetimes, although warming climates may open some of them up to road development.[21] But the world's remaining roadless areas also include some large areas of astonishing fertility and biodiversity. Indeed some of the most biodiverse regions on the planet remain relatively roadless: the great rainforests of the Amazon, central Africa, Borneo and New Guinea. More than 80 per cent of Brazil's surviving natural vegetation has no road or rail infrastructure. Many of these last extensive roadless areas lack any form of environmental protection, but in ecological terms they are vitally important. They preserve the natural movements of animals through migration and dispersal and enhance landscape connectivity by preventing protected areas from becoming isolated in a rising sea of traffication. They support particularly disturbance-sensitive species, which might have been driven out of smaller fragments, and they offer native species a final redoubt against the spread along roads of invasive species. They drive the world's weather systems, protect watersheds that deliver drinking water to millions of people and contribute disproportionally to the capture and storage of carbon.

Recognising the damage that can spread like gangrene from even a single narrow highway, conservationists are fighting hard to keep these final roadless bastions completely free of traffic for as long as possible, but the economic and political forces ranged against them are formidable. In Brazil, the upgrading and paving of the previously abandoned Highway BR-319, which links Porto Velho in the country's notorious 'arc of deforestation' to Manaus in relatively intact central Amazonia, threatens what Philip Fearnside and colleagues call 'a path with no return to a tipping point of self-degradation and the loss of Amazonia's vital biodiversity and climate-stabilization functions'. Much of the land along the highway has already been illegally grabbed and logged, and the usual requirement for an ecological impact assessment has been waived.

The story of BR-319 is a depressingly common one, and it will become depressingly more so. Economic development, we are often told, needs cars; the Chinese have a saying: 'to make money, you must first build roads'. Putting these words into practice, China is funding a vast infrastructure project called the Belt and Road Initiative (BRI), which will construct a network of new highways, placing even greater pressure on the world's shrinking roadless areas. Part foreign aid, part economic self-interest and part international relations exercise, the BRI will, by its planned completion in 2049, boost road, rail and sea transport across the whole of Eurasia, down the eastern side of Africa and elsewhere. It is perhaps the largest infrastructure project the world has ever seen. The BRI might increase the threats already faced by a quarter of all the world's endangered birds, mammals, reptiles and amphibians. Large, wide-ranging predators such as the tiger and the Asiatic cheetah are particularly vulnerable.

At least another 5 million km of new paved roads will be gouged into the planet's surface by 2050 and some estimates suggest the final figure may be closer to 25 million km, enough to circle the equator more than six hundred times. And the majority will be built in tropical regions that contain Earth's most diverse and fragile ecosystems.

Climate refugees

Roads kill huge numbers of animals that try to cross them, and lock those that don't cross into ever shrinking traffic islands. At the same time, they accelerate the spread of potentially damaging invasive species and open up new areas for human exploitation. These are already huge problems for our remaining wildlife, which is struggling to survive in a world full of other environmental woes, but they will become even more serious threats as our climate warms over the coming decades.

A species' distribution – the parts of the world where it lives – is determined by a host of complex interacting processes that we are still a long way from fully understanding. It is clearly tied to the types of habitats it can survive in, but habitats that appear suitable usually exist outside its range too. Reed warblers, for instance, are largely tied to reedbeds, but there are plenty of reedbeds in the world that have no reed warblers. Competition with other species may play an important part here – the presence of one species of reed warbler in a reedbed may exclude another. But perhaps the most important factor determining where a particular species lives is climate. Each of the world's species inhabits its own individual climate bubble, within which temperature, rainfall, humidity and so forth are all within the limits that it can tolerate. If changing conditions stretch these boundaries beyond the species' tolerance they have no choice but to move on, like swallows in autumn. We know from the fossil record that the response of most species during past periods of rapid climate change was not to stay put and try to tough it out, but to follow their moving climate bubbles wherever they led. Already we are seeing the same pattern starting to take place in our rapidly warming world. In the UK, the distributions of species adapted to cooler conditions are starting to move northwards and uphill, trying to keep pace with their cool climate bubbles, and those favouring warmer climates are taking their place in the southern lowlands. In fact, the distributions of our bird species are shifting in exactly the way that computer models developed two decades ago, using only climate data, predicted they would. But this essential mechanism for coping with a rapidly changing world will be increasingly thwarted by roads.

For a species to keep up as its climate bubble moves across the landscape, it needs to be able to spread into new areas as they become favourable. Otherwise it may get left behind, trapped within a new climate that it may not be able to tolerate. In an era of rapid climate change, wildlife needs landscapes to be permeable, allowing each species to adapt to changing conditions in the optimal way. For many species, and particularly for road-avoiders, our dense network of tarmac blockades will prove to be a significant problem. They will somehow have to find their way across roads, perhaps hundreds of them, or face eventual extinction. Some species, birds and bats for example, have the advantage of flight to carry them over major highways, but others, such as molluscs, amphibians and reptiles, will have to overcome their avoider tendencies when their climatic migration brings them to the sides of busy highways, or they will face certain extinction. Plants, too, will become increasingly threatened, particularly those whose seeds are not spread over long distances by wind or animals.

Some roads will prove more serious obstacles than others. In the UK, the ranges of many species are moving northwards in response to climate change, so major roads running from west to east (the M4, A14 and M62, for example) might prove to be more serious barriers than those running from north to south (the A1, M1, etc.). The latter, however, will help spread the wave of new invasive species, such as common ragweed, that climate change is certain to bring.

There is yet a further problem for species struggling to adapt to climate change. We have seen that one of the consequences of fragmenting our landscape into a patchwork of traffic islands is that populations of animals trapped within them lose their genetic diversity. With a loss of genetic diversity comes a loss of adaptability – there are fewer individuals within the population that might be adapted to survive and keep the population going if conditions change for the worse. Roads and their traffic are a significant source of the pollutants that cause climate change, and they also severely limit the ability of animals and plants to adapt to a warming world.

Climate change has turned our plants and animals into refugees, seeking a better life – or indeed any sort of life – elsewhere. To survive, they will need to break out of the traffic islands that constrain them and cross a hostile, heavily trafficated landscape. But in doing so they will face more threats than simply roadkill. For as well as splintering the landscape, road traffic has also shattered the soundscape.

CHAPTER 6

Thunder Road

A map of Britain's roads is, to all intents and purposes, a map of Britain's noise. Most of us live in cities, which reverberate to the din of traffic; around 80 per cent of urban noise comes from motor vehicles. But the noise of traffication is utterly pervasive; unless we live near an airport flight path or a busy railway line, road traffic is usually the dominant source of noise in our lives wherever we live. In every European country, whether in cities or in the countryside, the number of people exposed to the noise of road traffic far exceeds those exposed to other forms of noise pollution.

In 2016, the European Environment Agency published a beautiful report called *Quiet Areas in Europe*. It presents a map of Western Europe with its land shaded from cold blue, showing the quietest areas, through green and yellow (moderate noise) to orange and finally hot red, indicating the noisiest areas. What is striking about the map is just how much of Britain is polluted by noise. South of a line from the Severn to the Humber, an area that is home to around half of Britain's population, our dense road network coalesces into a more or less continuous blanket of angry red and orange clamour. There are very few remaining areas of quiet blue; if you want to wander in southern England in peace and quiet, far from the madding car, the map suggests that Dartmoor, Salisbury Plain, parts of the North Wessex Downs and the Brecks of East Anglia are the few options still left to you. If you have a feeling that you have come across these places already in this book, then you are quite right – they are the few remaining areas of any size in southern England that lie more than a couple of kilometres from the nearest road (see Map 4, p. 39).

North and west of the Severn–Humber line the country is generally quieter, particularly in the highlands of Wales, northern England and central Scotland, and when averaged out by area (though not by head of population) the UK is far from being the noisiest country in Western Europe. England on its own would be quite near the top of the list of Europe's noisiest countries (and Scotland quite near the bottom), but pity the poor Belgians, whose country is shaded almost entirely in red. Most of the Netherlands is shaded orange, not because it is their national colour but because it is their national noise level. Liechtenstein, Luxembourg and Malta are also destinations best avoided by those seeking car-free tranquillity. But few places on the continent are completely unpolluted by human-made clamour. The only really extensive areas of deep blue, free of any human noise, are the rugged roadless refuges of central Iceland, the high Alps and remote western Norway.

It is a fascinating thermal image of the continent's noise. But it is more than that – it's also a map of Europe's health problems.

'This is the source of our sickness'

The word *noise* imparts a degree of unpleasantness or intrusion, a sound that is unwelcome. Say it to yourself – a noise *annoys*. It is easily dismissed as a mere inconvenience, a nuisance at most, but transport noise is actually the second biggest environmental cause of ill health in Europe after air pollution (itself largely the product of road traffic).

It is no coincidence that the word *noise* comes down to us from the Latin *nausea*. We were warned about its dangers 2,000 years ago. Writing in Rome in around AD 100, the poet and satirist Juvenal saw the problem clearly:

> Insomnia is the biggest killer here ... Show me the
> apartment that lets you sleep! You have to be filthy rich
> to find rest in Rome. This is the source of our sickness.
> The wagons thundering past through those narrow,
> twisting streets, the oaths of draymen caught in a traffic
> jam, would rouse a dozing sea-cow – or an emperor.[22]

In 1905 the Nobel Prize-winning scientist Robert Koch, working in Berlin on horrific diseases such as tuberculosis and anthrax, predicted: 'The day will come when man will have to fight noise as inexorably as cholera and the plague.' That day arrived in Berlin rather sooner than he might have expected, because the city's famous Potsdamer Platz would soon become Europe's busiest traffic junction: in July 1928, a total of 33,037 vehicles crossed it in a single day.

Juvenal and Koch have been proved entirely right by recent medical research. We are slowly waking up to the fact that traffic noise is not simply annoying; it is extremely dangerous. The slight damage that prolonged exposure to road noise might do to our hearing pales into insignificance beside a host of deeper and far more life-threatening problems. Each year, more and more studies are being published on the damage it causes to the human body, and to the human mind.

The root of the problem is that we subconsciously perceive noise, even at fairly low levels, as a danger signal. Our bodies react involuntarily with a 'fight or flight' response, squirting stress hormones such as cortisol and adrenaline into our bloodstreams that trigger an increase in heart rate, blood pressure and anxiety. If this happens over prolonged periods, particularly at night when it disrupts vital sleep patterns, it can alter our body chemistry with devastating impacts on our health. People exposed to higher levels of road traffic noise are more likely to suffer from depression, bipolar disorder, anxiety, asthma, diabetes, stroke, coronary heart disease, chronic inflammation, brain damage and certain cancers. They drink more, smoke more and suffer higher rates of obesity than people living in less trafficated areas. Their children are more likely to be born prematurely and underweight, and as they grow up they are more likely to suffer from cognitive impairment, poor attention, slow development, behavioural problems and allergic conditions. A long-term study of thousands of female nurses in Denmark found that after other factors such as lifestyle and air pollution were accounted for, traffic noise pollution significantly reduced their life expectancy. Recent research indicates a worryingly strong association between road traffic noise and all forms of dementia, particularly Alzheimer's disease.

Around a quarter of all Western Europe's people live in areas where traffic noise levels are considered to be harmful to health. Long-term exposure to noise, mostly from roads, is estimated to account for 12,000 premature deaths, 43,000 hospitalisations, 48,000 new cases of heart disease and nearly a million new cases of hypertension (high blood pressure) in Europe each year. And each year sees the publication of more and more research that suggests that these numbers may be considerable underestimates. The World Health Organisation estimates that over a million healthy years of life are lost each year to road traffic noise in Western Europe alone, and this is predicted to rise to 1.7 million by 2030. More than 100 million Americans are similarly exposed to dangerous levels of noise, most of it from roads.

Traffic noise is not just an annoyance; it is a profoundly important health pandemic whose severity we are only slowly recognising. Neither is it a problem that we can simply get used to; our bodies and minds do not acclimatise to noise the longer we are exposed to it. In fact, the opposite may be true – long-term exposure to the din of traffic might actually make people *more* sensitive to its effects. Even while you sleep your body continues to respond chemically to the alarm your subconscious nervous system triggers when vehicles pass by outside, damaging your health without even waking you up. Roads produce a roar of attrition, insidiously eroding our health day by day, night by night.

There is a huge disparity between the scale of the health risks posed by road traffic noise and our willingness to tackle it. Our casual attitude towards the health impacts of traffic noise recalls the way that we quietly looked the other way when it became clear in the 1950s that smoking tobacco is a killer. The evidence is scientifically overwhelming but socially and economically unwelcome, and we cannot identify victims of road noise in the same unambiguous way we can identify those killed in accidents. Public opinion is nowhere near gathering the momentum required to force big, brave political changes on the issue, but things are slowly changing. In 2022, the American Public Health Association changed its definition of noise from 'unwanted or undesired sound' to 'unwanted and/or harmful sound' to recognise its serious health impacts.

There are legal limits to the noise an individual vehicle's engine can make, but there is no law to limit the combined din of traffic. Motorists actually enjoy legal protection in Britain in this regard – you can't sue noisy drivers for ruining your health, and local authority powers to deal with noise problems explicitly exclude road traffic. European Union law on this issue requires Member States to do little more than simply map and monitor the problem. Yet a single unsilenced motorbike crossing a city at night has been estimated to disturb (*injure* might be a better word, given the health impacts) over 11,000 people. Even the ancient Romans enjoyed better legal protection from traffic noise than we do today; Julius Caesar introduced laws to restrict the driving of chariots and wagons in towns and cities, largely because of the loud clatter of their wheels on the stone streets.[23] The transition from internal combustion engine to electric motor over the coming few decades might, at best, slightly reduce the problem of traffic noise, but equally it could even exacerbate it, as we will see in a later chapter. Road noise is a problem that will be with us for a while to come.

Behind all the health problems lies, inevitably, social inequality. Areas of high road noise tend to be areas of low house prices and lower car ownership, so the burden of health costs falls disproportionately on the less affluent sectors of society, those who benefit least from road travel. Londoners suffering the highest levels of traffic noise are those with the lowest rates of car ownership. The dangers of passive motoring have not yet percolated into our collective conscience to the extent that the dangers of passive smoking eventually did, but many public health experts have drawn clear parallels between them.

The stress of roadside living

Other animals suffer exactly the same sorts of health impacts from road noise as we do. A huge amount of research, from both the field and the laboratory, has shown that animals exposed to vehicle noise suffer higher stress levels and weakened immune systems, leading to disrupted sleep patterns and a drop in cognitive performance. They suffer an impaired ability to learn, solve problems or remember where things are, failings that in the brutally competitive world of wild nature usually mean the difference between life and death. When exposed to prolonged traffic noise, zebra finches become less good at finding food, roosting bats become more susceptible to disease, great tits suffer severe sleep disruption and frogs become unable to produce antimicrobial proteins to fight infection. Mice develop problems with motor coordination and their brains shrink. The heterophil to lymphocyte blood cell ratio, a good indicator of stress levels, is elevated in tropical forest birds exposed to high levels of traffic noise. There are many other examples of such effects, across a wide range of animals. Even aquatic animals are not exempt, since road noise can reverberate through bridges and culverts and into lakes and rivers. Just like terrestrial animals, fish and the tadpoles of frogs and toads become stressed and change their behaviour in the presence of road noise.

As with people, these symptoms can arise even at low levels of disturbance. One of the first studies to show that traffic noise can elevate stress levels in birds was undertaken in Gunnison National Forest, Colorado, where white-crowned sparrows nesting near a road were found to have higher concentrations of the stress hormone corticosterone in their blood than those nesting further away. This may have explained why birds breeding near the road more often abandoned their nests. What was telling about this study was that the road in question was an unpaved track carrying only a few slow-moving vehicles each day. Another American study, this time of breeding tree swallows, found that exposing breeding birds to just six hours of artificial traffic noise every second day caused their chicks to grow more slowly and to be in worse health than those left in peace and quiet. This appears to be a common pattern; the stress caused to birds by even low levels of traffic noise results in a drop in the number of eggs laid and the health of the chicks that hatch. A study using data on over 100 bird species from across the USA has found that as traffic noise increases, so nesting success falls. Noise-induced stress affects not only the health of individual birds, but threatens the survival of entire populations by reducing the number of new recruits being produced each year.

The same is true with mammals. Researchers working along some quiet roads in a national park in Spain collected roe deer droppings and analysed them in the laboratory for traces of stress hormones. Even though traffic was low on all the roads, roe deer near the slightly busier roads had higher levels of stress hormones than those near the quietest roads. Very similar results have been found in studies of impala in Tanzania's famous Serengeti reserve, and wood mice in Spain.

It really does not take very much traffic noise at all for animals to start to suffer physical harm. Levels of noise that even the most sensitive of us might find entirely tolerable can have severe impacts on wildlife. And as with humans, there is little evidence that the impacts of traffic noise on wildlife diminish through long habituation, and may indeed become worse: the stress response of tree swallows has been shown to increase the longer the birds are exposed to the sound of traffic.

Road noise can also impact the health of animals indirectly, for instance by reducing their hunting efficiency. Owls and bats, which rely heavily on their acute hearing to find prey, struggle to locate food in conditions of high road noise; one study suggested that bats showed a two- or threefold reduction in feeding efficiency even at distances of over 600 m from a road.

The impacts of road noise can manifest themselves in some unexpected ways. In most species of frog the male has a vocal sac, a flexible membrane of skin under the throat that is pumped up with air like a balloon when he is giving his mating calls. The sac serves two purposes – it amplifies the call, and it also forms part of the frog's amorous display, adding a visual element to his acoustic advertisement. Vocal sacs are therefore often brightly coloured to maximise the effect. European tree frogs exposed to traffic noise show elevated levels of stress hormones in their blood and weakened immune systems, as in many other animals. But these changes have the further effect of causing their bright orange vocal sac to lose its colour. Moreover, it is the fittest males, which have the brightest vocal sacs to start with, that are most affected. This loss of colour means that females struggle to pick out the best mates, giving the less fit males a better chance of passing on their genes to the detriment of the next generation.

Most worrying of all is a growing body of evidence showing that noise can affect the health of wild animals at a genetic level. Our chromosomes, and those of all other animals, are protected at their tips by a cap called a telomere. Rather like the little plastic sheathes at the ends of your shoelaces (called aglets, apparently), telomeres stop DNA strands from fraying at their tips. Telomeres get shorter each time a cell divides and eventually they get too short to do their job, so our cells start to deteriorate and eventually stop functioning properly. Our bodies age, and the incidence of diseases such as cancer starts to rise. So the longer your telomeres are to begin with, the better your chances in life. The length of an animal's telomeres appears to be a good predictor of how long it will live, more violent causes of death excepted, and we know that they can be shortened by stress.

In 2015, a team of French scientists made an important discovery. They found that house sparrow nestlings experimentally exposed to road traffic noise had significantly shorter telomeres than those raised in quiet conditions. The following year a Swedish research team reported a similar result, showing that young great tits reared in urban environments have shorter telomeres than those in quiet rural areas. Rather neatly, they proved that it was the environment itself that was responsible, rather than the health or quality of the parents, by swapping chicks between nests in urban and rural habitats. Similar results have since been found in zebra finches, blackbirds and tree swallows. Birds raised in the presence of traffic noise are prematurely aged, and their future lifespans already curtailed, before they have even left the nest. Remarkably, we know more about the effects of traffic noise on telomeres in birds than in our own species; I have been unable to find a single published scientific study of the effects of road noise on telomere length in humans.

The weird world of decibels

When measured at the kerb, road traffic generally produces somewhere between 60 and 85 decibels (dB) of noise. Human health starts to suffer when traffic noise levels exceed a daily average of 53 dB, and at night anything over 45 dB is considered to be potentially harmful. Communication between birds can start to break down at 50 dB (the level of noise that people start to find annoying), and some animals show adverse effects if road traffic noise exceeds just 40 dB, less than that in the average library.

But what do all these numbers actually mean? Before we can start to understand the environmental problems of noise, we need to address the slightly tricky issue of how it is measured.

Put simply, decibels measure the energy (or, more technically, the *amplitude*) of a sound. The name has nothing to do with ringing bells; it comes from the inventor of the telephone, Alexander Graham Bell (the second 'l' got lost somewhere along the way). Decibels are difficult to interpret because they follow a logarithmic scale: every additional 10 dB represents a tenfold increase in sound energy (which is not *quite* the same as loudness, as we shall see). Therefore, a sound of 70 dB contains ten times more energy than a sound of 60 dB, a hundred times more than a sound of 50 dB and a thousand times more than a sound of 40 dB. The benefit of squeezing numbers up like this is that the extraordinary range of sounds our ears can register, from the faintest hum of a flying gnat to a trillion-times-louder firework display, can be conveniently captured on a scale that runs from 0 to 140 dB (anything above that will quickly deafen us).[24]

In the weird way of logarithms, small differences in decibels mean big differences in sound. A useful rule of thumb is that each increase of 3 dB represents a *doubling* of sound energy. If you are running a washing machine that puts out 70 dB of noise and you then switch on a vacuum cleaner that also produces 70 dB, the combined din in the room measures a rather unintuitive 73 dB. It is certainly not 140 dB, the quite literally deafening racket of 10,000,000 washing machines.[25]

There's a slight complication here, which is that decibels measure all the different parts of a noise, from the lowest frequencies to the highest, whereas human ears hear only some of them. Many sounds are too low-pitched (infrasound) or too high-pitched (ultrasound) for our ears to pick up at all. Even within the range of frequencies that our ears can detect there is a lot of sound they miss, particularly at lower frequencies: a double bass sounds quieter to our ears than a trumpet, even if the two instruments register the same number of decibels.

What this means is that while a sound of 70 dB is *technically* ten times louder than a sound of 60 dB, to human ears it sounds only about twice as loud, and four times louder than a sound of 50 dB. A corrected measure, called dB(A), is used to measure loudness as the human ear perceives it, but this does not help us to understand how non-human animals experience sound.

There's one final little quirk about decibels. The faintest sound that a healthy young person's ear can hear is set to a value of zero dB, but the sharper ears of other animals can detect sounds quieter than this. These super-quiet sounds can still be measured on the decibel scale, but they get negative values: a sound ten times fainter than anything we can hear registers -10 dB. Cats, dogs and owls may be able to hear sounds as faint as -20 dB, a hundred times quieter than anything human ears can detect.

Soundscape pollution

The harmful cacophony produced by road traffic is one of the biggest contributors to a growing problem generally referred to as noise pollution. But *noise pollution* is not a very logical description of the problem. Air pollution is the pollution of air, water pollution is the pollution of water, and soil pollution is the pollution of soil. But noise pollution is not the pollution of noise; rather, noise *is* the pollution. This might sound like a pedantic quibble, but it raises an important question: what is it that noise actually pollutes? It can't be pure silence, otherwise the song of a nightingale or the gentle lapping of waves on a beach would also be classed as noise pollution. The din of road traffic sullies something far more complex, subtle and biologically important than silence. Let me propose an alternative name for the problem as it applies to wildlife: *soundscape pollution*.

A soundscape is the acoustic environment that animals inhabit. It is made up of the wide range sounds produced by other animals (called *biophony*) and an even wider range of non-biological but nevertheless natural sounds, such as the trickle of a waterfall, the sigh of the wind in the willows or a distant rumble of thunder (*geophony*). To these natural soundscapes are now added the unnatural sounds of our own activities, foremost among them traffic noise. This has been given the convoluted name of *anthropophony*, but it is easier to call it soundscape pollution.

Each animal inhabits its own unique natural soundscape, different even from that of other animals living in the same place because each has its own individual ability to detect and interpret the sounds around it. The same faint rustle in the undergrowth might signal food to one animal, a mating opportunity to a second and life-threatening danger to a third.

Hearing did not evolve as a way of detecting the communications of other animals. It clearly evolved long before singing or barking or chirping came along, because why sing or bark or chirp if nothing else can hear you? Instead, the driving force behind the evolution of hearing was the wealth of information it gives an animal about its environment, the ability to 'see' beyond its eyes, to detect the snap of a twig broken by a stalking leopard, the movements of a mouse under snow, the distant crackle of a bushfire. An animal that is able to hear and interpret a soundscape may receive a much wider range of information about its surrounding environment than it gets from its other senses – eyes can generally look only in one direction and need at least some light to work, whereas ears can detect sounds from all around even in total darkness.

The ability to hear sounds evolved originally in aquatic life to gather information about the surrounding environment in murky waters; the remarkably complex structure of our ears is evolution's solution to the problem of bringing a fluid-based sound detection system onto dry land. Once hearing arrived, it unlocked the extraordinary potential of acoustic signalling – the deliberate exchange of information between animals through sound. Sound is used to convey information to members of the same species, about breeding condition or territorial boundaries, for example, or to communicate information to other species – the warning hiss of a snake or rattle of a porcupine's quills. As always in nature there are exceptions – bats, sperm whales, oilbirds and some other animals produce sound to find prey or to navigate in the dark – but most of the sounds created by animals are designed to say something explicit to other animals. They say 'come here', 'stay away', 'watch out', 'stick together', 'food here', 'fight me', 'feed me', 'mate with me'. These messages are not merely incidental to animal life; they are essential to it. Without them, natural ecosystems will rapidly unravel and break down. The dawn chorus of birdsong is not, as Victorian parsons would have it, a daily concert put on for our delectation, but a desperate acoustic fight for survival in an unforgiving world.

Human domination of the planet was achieved largely through our ability to communicate complex information through speech, but our unique technological genius now allows us to feed ourselves, protect ourselves and exchange vital information without needing vocal communication at all. In the digital age, less and less of the information exchanged between us requires the uttering of words or the hearing of ears; we increasingly communicate through writing, keyboards, screens and signs. But in the red-toothed world of wild nature, any animal whose soundscape is compromised faces a huge range of new threats to what is already a precarious existence.

The delicate soundscape that animals have evolved within, and contributed to, over millions of years, and from which they extract so much vital information, is as beautiful and complex to the ear as any landscape is to the eye. It is a subtle whispering gallery of faint swishes and little taps, an acoustic gossamer of minus-decibel squeaks and crackles, each conveying vital information to those with ears to hear them.

And it is utterly obliterated by 80 dB of traffic noise.

Singing on Thunder Road

Of all the acoustic communication systems that have evolved in nature, birdsong is surely the most conspicuous and the most complex. Birds, cetaceans (whales and dolphins) and humans are the only animals that have an extended period of learning when it comes to making meaningful sounds, and the only ones to use complex sounds in communication.[26] The evolution of extremely complicated yet gloriously beautiful songs, and an equally dazzling array of calls and other vocalisations, might suggest that birds have very acute hearing, the better to detect these subtle but information-rich sounds. It comes as a surprise to most people, then, to learn that the hearing of most birds is less good than our own. This is largely because we have a large external ear to collect and focus sounds and a much wider auditory canal to channel it towards the sensory cells of the inner ear. Whereas we (and all other mammals) have three small bones in our middle ear that articulate to conduct the vibrations of the eardrum to the inner ear, birds have only one. This probably explains why we can hear sounds that are both higher- and lower-pitched than most birds can. Most birds can only hear (and make) sounds up to around 8 to 10 kHz, whereas the ears of a young person can hear the clicking of bats at over 20 kHz.[27]

More surprising, perhaps, is that we can also hear *quieter* sounds than most birds can. The quietest sound that a person can hear is, as we have seen, 0 dB, but many birds appear to be unable to detect sounds below 10 or even 20 dB. Of course some specialised acoustic hunters, such as owls, have better hearing than we do, and some birds might have more acute hearing than ours in certain narrow bands of pitch, but the average human ear outperforms the average avian ear in pretty much every regard.

This means that in a noisy roadside environment, a person listening to birdsong will greatly underestimate the effect the noise is having on bird communication; a person can hear birds singing or calling at around twice the distance that the birds can hear each other. Birds therefore have a bigger problem when it comes to picking out important sounds above the background rumble of traffic than the evidence of our own ears might suggest.[28] Their soundscape is heavily polluted, and this can start to affect them from the day they hatch from the egg.

As with children, young birds are particularly vulnerable to noise because it interferes with learning at a critical stage of their development. Many birds learn their song from adults of the same species early in life. In zebra finches, the song-learning phase starts at around 25 days after hatching, when the chicks start to memorise the song they hear.[29] Around 35 days after hatching, they begin to develop their own songs, gradually matching its structure to the remembered adult song. At the age of around 90 days their song *crystallises*; in other words it stops changing and becomes fixed – it is the song the bird will sing for the rest of its life. Experiments have found that zebra finch chicks exposed to real-world levels of traffic noise take longer to learn their songs, and those songs take longer to crystallise. Moreover, their final crystallised songs are much less accurate copies of the parental song than those of birds raised without traffic noise. This is probably because (as shown by a different study) the regions of the avian brain that are involved in song-learning are smaller in birds exposed to traffic noise than they are in birds raised in undisturbed conditions, presumably as a result of increased stress.

What this means is that in noisy environments, badly learned versions of the original songs will be badly learned by the next generation and so forth until, as in the party game Chinese Whispers (or Telephone in the USA), all the meaning contained in the original ancestral song has been lost. This raises the possibility that birds breeding near roads will, over time, become increasingly unrecognisable to other members of their own species. Just as animals are divided physically and genetically by roads, so road noise causes populations to start to drift apart and fragment acoustically.

Song is not the only way that birds communicate vocally. The tits (or titmice and chickadees if you live in North America) are brainy, highly social birds that seem to be particularly good at spotting approaching danger, particularly in the form of birds of prey. When they see danger scudding in they give a high-pitched alarm call to alert their comrades, telling them to hide or scatter. Many other species of birds have learned to recognise these alarm calls, and know that when a great tit or a black-capped chickadee or a tufted titmouse gives an alarm call, they also need to take evasive action. It is an anti-predator communication network that many bird communities rely on. But it is fatally disrupted by traffic noise. Experiments have shown that traffic drowns out these alarm calls, making birds more susceptible to being killed by predators. Mammals too suffer from this problem. In Chapter 5 we met the peculiar and endangered Stephens' kangaroo rat. This species does not call its warning when danger approaches, but instead stomps on the ground with its outsized back feet to alert its comrades, as rabbits do. Unfortunately, the sound frequency of its thumps is very similar to that of traffic noise. This has two consequences: first, its alarm stomps are drowned out by traffic noise so the warning message is not passed on; second, traffic noise itself sounds like an alarm stomp, so animals living near roads exist in a perpetual state of anxiety.

To compensate for the drowning out of their natural alarm systems, animals living near roads become more vigilant, spending more of their time looking around for danger and consequently having less time to feed. In Canada, elk change their behaviour from feeding to vigilance near roads carrying as little as one vehicle every two hours. Clever experiments carried out in the USA showed that when traffic noise is played through a long line of loudspeakers in otherwise undisturbed woodland, creating a 'phantom road' comprising only sound, the birds that were not scared away entirely stopped putting on weight and suffered a fall in body condition, presumably because their feeding rate had fallen. Similar experiments that played recordings of traffic noise near colonies of American prairie dogs and dwarf mongooses, both gregarious mammals that use warning calls to alert family members to the presence of danger, found exactly the same thing: they became warier, and spent more time looking around for predators and less time feeding. Road noise creates what has been called a 'landscape of fear', and animals suffer as a consequence.

Researchers studying the songs and calls of birds, frogs and prairie dogs living near roads have also noticed something else, something rather strange and harder to explain: traffic noise causes animals' voices to change.

Going piccolo

You are talking to a friend in a bar. It is early in the evening and still fairly quiet so you chat away in your normal voices; each of you hears all the meanings and intonations in everything the other says. But as the bar starts to fill up and the background chatter gets louder, you begin to lose the thread of what your friend is saying, and vice versa. What you say depends for its full meaning not only on the words you choose but also on the subtle nuances of how you say them, so your communication may start to break down at even fairly low levels of background noise. You may still catch most of your friend's words, but the precision of the information exchanged between you starts to blur, and errors and misunderstandings start to creep into your conversation. Your pleasant tête-à-tête has become disrupted by what is technically known as a low signal-to-noise ratio: the information that you are trying to communicate to your friend (the signal) has become so compromised by background noise that a critical part of it is lost. Your conversation increasingly starts to resemble an acoustic version of a heavily redacted document, with key words or even whole sentences being censored out.

The first thing you will do as the bar starts filling up, almost without knowing you are doing it, is to try to improve your signal-to-noise ratio by changing your behaviour. You can do this by making sure your mouth is turned towards your friend while you speak, and your friend might turn their head towards you while listening to catch your words. You may both sit forward so that your heads are closer together, and your lip movements, facial expressions and gesticulations become exaggerated to help clarify and reinforce the information in your words.[30] You may find yourselves adopting a simpler vocabulary, favouring shorter words over longer ones.

But this will only work so far. As the chatter around you becomes louder, you will be forced to put more energy into your speech. This triggers a number of interesting processes that are collectively known as the Lombard effect. The way you physically produce sound will change – you use greater lung capacity, for instance, your vocal cords and glottis move differently and your mouth opens wider. Your vowels become longer and you stretch out the more informative words when you say them, while shortening the less important ones. The amplitude, or loudness, of your voice increases – you progress from talk to shout to yell in order to keep your conversation going above the rising racket. As you do so, the pitch of your voice inevitably rises: try yelling or screaming in a deep voice; it feels very unnatural. Professional singers are well aware of this, and train their voices so that when they need to sing louder they are able to correct their natural tendency also to sing higher.

The extent to which your conversation is disrupted will depend on who is chatting around you. If you and your friend are women and the people talking around you are all deeper-voiced men, you will find it easier to hear each other than you would if the background chatter was that of other women. This is because against a background hum of deeper voices, you face less competition in the sound frequencies you are using. In the same way, it is easier to pick out the sound of the piccolos in a loud orchestral finale than that of the violas: few other instruments reach the high-pitched notes of the piccolo, so there is nothing to mask them – they have the highest parts of the music all to themselves. The viola, on the other hand, overlaps the ranges of both the cello below it and the violin above it, so its sound is more camouflaged. And any master of ceremonies knows that the best way to attract attention in a roomful of loud conversation is not to try to shout over the competition, but to tap a spoon against a wine glass.

Animals for which acoustic communication is a vitally important part of life, whether territorial birds, courting frogs, chirping grasshoppers or alarm-barking prairie dogs, are sometimes able to make themselves heard against the din of road traffic in similar ways. Up to a certain level of background noise they may change their behaviour to compensate, as we do. Urban birds, for example, sing more frequently at night in comparison with their rural cousins, taking advantage of the lower levels of traffic noise in the small hours. Some may give up singing altogether during the morning and evening rush hours. Grasshoppers chirping on road verges reduce their call rate in response to each individual passing car, beginning again when it has passed. Robins singing in noise-polluted areas often choose higher perches from which to sing in an effort to broadcast their song further (even if it places them at greater risk of sparrowhawk attack).

If behavioural changes fail, however, many species of birds (though certainly not all) are able to raise their voices and sing or call louder. Owing to the Lombard effect, the pitch of their voices also rises – they sing both louder and higher. Scientists are still not able to agree which of these two changes is the more important when it comes to overcoming background noise. Some argue that by raising the pitch of their voice birds are able to 'go piccolo', so their song competes less with traffic noise and can be picked out more easily by those for whom it is intended – other birds. Most road noise is low-pitched – we talk about the *rumble* or *thunder* of traffic, not the *whistle* or *squeak* of it – but some road traffic noise falls in the higher frequencies of sound at which birds usually sing. So perhaps by raising the pitch of their song, birds are trying to climb above the soundscape pollution. Even chicks in the nest appear to respond in this way, raising the pitch of their food-begging calls to their parents in nests near roads. Other scientists, however, consider that singing louder is a much more effective response to soundscape pollution than singing higher, and that the rise in pitch is nothing more than an inevitable consequence of the Lombard effect.

Either way (or both), singing or calling birds try to counter the effects of road noise by raising the volume and pitch of their song. But this does not necessarily mean that they have escaped the problem of soundscape pollution, because in adapting their voices to be heard over the din of traffic, the quality of their song is reduced. Song in birds serves to secure a mate and to defend a territory, both of them essential requirements. We know very little about which particular qualities of the complex songs of birds it is that females find attractive, or rival males find intimidating, but several studies have shown that when a male alters his song in the presence of traffic noise, females and rival males – the very birds that he is trying to influence – respond less strongly towards it. In many species, females appear to prefer males that produce lower-pitched songs, the exact opposite to what traffic noise does to their prospective suitor's voice. Some species sing less complex songs in noisy environments, either because birds find it hard to sing songs that are both loud *and* complex or because complexity is lost in the din of traffic and so it is not worth the effort of creating in the first place. So males that can change their songs in the presence of road noise might be more easily heard by other birds than those that cannot, but their songs may have less impact as a result. We know almost nothing about how the changes that birds make to their songs in the presence of traffic noise impacts their ability to find a mate or hold a territory, but anything that causes them to change this most vital of signals from its natural pattern surely cannot be anything other than a problem.

Birds are not the only animals to change their voices in the presence of traffic noise. Some frogs (but again, not all) show an equally rapid response to soundscape pollution, by similarly raising the volume and pitch of their advertising calls. The louder the traffic noise, the louder and higher the frogs call.[31] But as with birds, the modest benefit they gain by raising their voices comes at a cost, because female frogs, like female birds, appear to prefer males with lower-pitched calls.

Mammals, too, change their calls in the presence of road traffic noise, and generally in the same way as birds and frogs do – by increasing loudness and pitch. We have seen already that prairie dogs do this in their alarm calls. California ground squirrels near noisy highways also change their alarm calls by shifting acoustic energy into sounds that overlap less with the sound of traffic. Bats increase the loudness and reduce the complexity of their social, non-hunting calls near roads, and when hunting they change the rate at which they emit their pulses of moth-seeking sonar. Even grasshoppers, crickets and cicadas, which produce sound in a completely different way to vertebrates, respond to traffic noise by raising the pitch of their chirpings. These shifts in vocalisation, whether in birds, frogs, mammals or insects, can start immediately the traffic noise begins. If you broadcast artificial traffic noise to frogs in a normally quiet environment, they almost instantly respond with an increase in call pitch. And in rare cases where traffic noise ceases, animal vocalisations can return to normal just as quickly.

In the unprecedented spring of 2020, the COVID-19 pandemic forced countries around the world to restrict the movements of people in order to slow the spread of the disease. This led to the first significant drop in road travel since the start of mass motoring, and all around the world cities and busy roads fell quiet. This lockdown offered scientists a unique opportunity to see how animals respond to newly unpolluted soundscapes. In San Francisco, a team of researchers studying the impacts of traffic noise on the songs of white-crowned sparrows made an interesting discovery. Their previous work had shown, in line with many other studies, that sparrows in the polluted soundscapes of San Francisco sing louder, higher-pitched and less complex songs than those just over the Golden Gate Bridge in quiet, rural Marin County. These are not just subtle differences that are detectable only by using complex scientific equipment: they sound utterly different even to an unaided and inexperienced human ear. But as lockdown was imposed, and traffic noise fell for the first time in over a century, soundscape pollution in the territories of urban white-crowned sparrows fell to levels no higher than in those of rural birds. In this fleeting interlude of peace, the urban sparrows immediately reverted to their natural song – quieter, lower and more complex, exactly like that of their rural cousins. Despite being quieter, their songs carried more than twice as far in the newly hushed air than their much louder songs had done against a background of road noise, showing that singing louder to beat the traffic has only a limited effect.

The study's lead scientist, Elizabeth Derryberry, succinctly summarised the effects of this brief outbreak of calm on her urban sparrows: 'They're not shouting anymore.'

Distant thunder

Road noise undermines the health of wild animals, it reduces the information they receive about their environment, and hence their ability to attract a mate or find food (or to avoid becoming someone else's food), and it interrupts vital communication networks. All of these impacts have potentially catastrophic consequences and, as we shall see later, the response of many animals is simply to abandon areas near roads.

But if animals only suffer from soundscape pollution within the noisiest first few metres from the roadside, then perhaps it would not represent too much of a problem for them in the overall scheme of things. After all, roads and their verges only cover one hundredth of Britain's land area, and animals living right next to them might be more at risk from roadkill than from noise. But if we look back at Maps 1–4 (pp. 36–39), and consider again the third element of traffication – pervasiveness – then we soon start to see that soundscape pollution from road traffic does not have to seep very far at all into the surrounding landscape before it starts to affect huge areas. Recall that anything that spreads even just 100 m from the roadside is already reaching a fifth of all Britain's land, and a third of England's. And we all know that we can hear the roar of traffic from much further away than that.

To assess the scale of the problem of soundscape pollution, we require two pieces of information: we need to know how far road noise carries into the surrounding countryside, and we need to know what level of noise starts to cause problems for wildlife. Perhaps predictably neither of them is easy to measure, but let's give it a go.

The total amount of noise produced by a road is determined largely by two things: the number of vehicles it carries and, perhaps even more important, the speed they travel. A car travelling at 70 mph makes twice as much noise as it does at 60 mph and an astonishing 25 times more noise than when travelling at 30 mph. The texture of the road surface is also important. Most of the noise produced by moving traffic comes not from the vehicles' engines but from the interaction of tyres and the road surface (called *rolling noise*).[32] This is why electric vehicles travelling at speeds over around 20 mph make almost as much noise as petrol-powered vehicles. The amount of rolling noise produced by a moving vehicle depends in large part on the road surface itself: newer, smoother roads produce less rolling noise than older, more patched-up ones, and some surfacing materials are inherently quieter than others. Wet roads generate far more noise than dry ones.

When measured at the edge of the tarmac, road noise usually registers somewhere between 60 and 85 dB. Of course, how much of this noise actually reaches our ears, or the ears of a deer or an owl or a bat, depends on how far we are from the road. If we get far enough away, a kilometre or more in the case of a busy road, we will eventually stop hearing it altogether (in heavily trafficated Britain, this is usually because in moving away from one road we have moved closer to another, whose noise drowns out the first). But distance is not the only factor that determined how much noise we hear from a road. The weather also plays an important role in how far sound travels, as you have probably noticed yourself. In warm weather, rising air carries sound upwards, dampening the roar of traffic, whereas cool, sinking air can push road noise downwards and outwards for long distances. Wind direction is also important, helping or hindering the spread of noise outwards from the road. Trees, buildings and other barriers, even certain soil types, can absorb and deaden road noise, or reflect it back across the road to increase the din on the other side.

It is therefore very difficult to work out how far traffic noise, with all its harmful impacts on people and wildlife, spreads outwards from any particular road and into surrounding areas. But we can at least make some realistic predictions based upon the mathematical properties of sound. Figure 5 shows how noise might plausibly spread outwards from each of three roads – a very busy road (carrying 2,000 cars per hour, and a tenth that number of light vans and heavy trucks), a moderately busy road (with half the traffic of the first) and a fairly quiet road (with a tenth the traffic of the first). The graph shows that the busiest of the three roads (the solid line) produces just over 80 dB of noise when measured at the roadside, falling

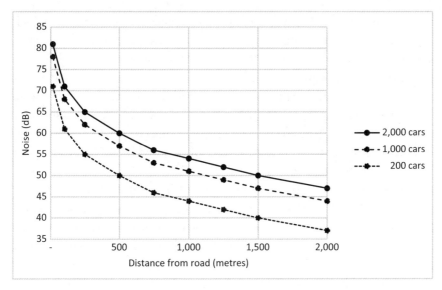

Figure 5. An illustration of how the noise produced by traffic falls away with distance from a road. Examples of three roads are shown: a dual carriageway carrying 2,000 cars, 200 light vans and 200 large trucks per hour (solid line), a busy A-road carrying half that (dashed line), and a moderately busy B-road carrying a tenth the traffic of the first (dotted line). In this simulation, vehicle speed was set at 60 mph. Noise from other nearby roads is excluded.

to 54 dB 1 km away and 47 dB 2 km away. A drop in noise from 81 dB at the roadside to 47 dB 2 km away might not sound like much of a fall over such a long distance, but we need to remind ourselves of the strange scale of decibels: a noise of 81 dB contains 2,500 times more sound energy than a noise of 47 dB.

If we know what level of noise creates a problem for our wildlife, we can use this graph to estimate how far away from each road its effects will be felt. Thankfully, we do have a little information on this. For some particularly sensitive bird species, we know that vocal communication can start to be disrupted in noise of 45 dB (which is also, according to the WHO, the safe average traffic noise limit for people at night), although the average across bird species is likely to be nearer to 55 dB.[33]

If we find 45 dB on the vertical axis of the graph and read across, we find that the most sensitive species will be affected 2 km or more from the two busier roads, and even the quietest road will cause problems for these species up to a kilometre away. Species whose communication starts to become affected when noise reaches 55 dB will be impacted if they are within a kilometre of the busiest road, or within 250 m of the quietest.

All these numbers now start to coalesce into a startling realisation. If we make the rather conservative assumption that an average road affects an average bird up to 500 m away, and if we look again at Map 2 (p. 37), which shows how much of Britain's land falls within half a kilometre of a road, we can see that across half of Britain, and three-quarters of England, the soundscape is damagingly polluted by traffic noise. This means that across at least half of Britain, and most of England, our birds are struggling to make themselves heard, and to hear other birds, because of the noise of our vehicles. The soundscapes of more sensitive species, which are affected a kilometre or more from a road, have been polluted by traffic noise across pretty much the entire country.

Other groups of animals show signs of being disturbed by noise as low as 40 dB, and a fifth of all studies that have examined the issue report detrimental impacts on wildlife of noise below 50 dB, so it is not just birds that are suffering from widespread, even ubiquitous, soundscape pollution.

Noise is an invisible form of environmental damage that seriously degrades wildlife habitats that are far away from the roads that produce it. There is a growing consensus among road ecologists that soundscape pollution is one of the most damaging environmental impacts of traffication, perhaps outstripping even roadkill in its impacts. Animals are vulnerable to roadkill only on those relatively few occasions they actually attempt a crossing, but they are affected by soundscape pollution all the time, even when far away from the road. For some species, practically our entire countryside has been degraded by this sonic doom. Its potential to damage wildlife over vast areas is no less than that of the industrialisation of agriculture or climate change. We will return to this issue in Chapter 8.

Fight or flight in the plant world

There is much we still do not understand about how road traffic noise impacts our wildlife populations. Our knowledge of how different sound levels affect animal behaviour is still in its infancy. We do not know, for instance, whether a constant low-level hum of traffic is more or less damaging than the occasional passing of very loud vehicles. Nor do we know much about how traffic noise might impact non-animal wildlife. Anything that impacts seed-spreaders, such as birds or small mammals, or pollinators, such as insects, could clearly influence our flora as well as our fauna with long-term effects on ecosystem structure and diversity.

It may seem inconceivable that road noise could influence plants directly, which probably explains why almost nobody has looked at the issue. Yet sound waves are known to interfere with the complex biochemical processes taking place within plants in a number of ways.

Recent research undertaken by an enterprising student at a high school in Ohio looked at the growth rates of some basil seedlings in a laboratory and found that plants grown in the presence of artificial traffic noise grew to little more than half the height and weight of those grown with a background of quieter natural sounds. This could easily be dismissed as the outcome of a one-off experiment, perhaps flawed by youthful inexperience, but the following year a team of experienced Iranian scientists, apparently unaware of the Ohio study, undertook a very similar laboratory experiment and got exactly the same results. They found that plants (scarlet sage and a type of marigold in this case) grown in the presence of artificial traffic noise grew more slowly than those grown in silent but otherwise identical conditions. Furthermore, they found that the balance of plant chemicals was radically altered by traffic noise – growth hormones were replaced by stress hormones and the production of toxic chemicals increased. Although the mechanism behind this response has yet to be explained, it may be very similar to the response of animals to traffic noise. Many animals (ourselves included) exposed to traffic noise produce stress hormones that trigger a fight-or-flight response. Clearly, plants don't have a great deal of choice when it comes to flight, but they do have some options when it comes to fighting. Many plants are able to produce toxic or distasteful chemical defences to protect themselves from predators such as caterpillars or grazing animals, but these are costly to produce and often do the plants themselves no good, so many plants don't produce them until they are needed. The tiny vibrations caused by sound waves from traffic noise may convince a plant that it is under attack from hungry insects, and stimulate it to invest more of its resources into producing defensive toxins and consequently less into growing. Plants, like animals, appear to become physiologically stressed by traffic noise, and this in turn affects their health and growth. If these early results are corroborated by further research, all sorts of important questions spring to mind - not least about how road traffic noise might be affecting the health and yields of the agricultural crops we all rely on for our survival.

This circles back to a final consideration about the important issue of our own health. Just as road noise takes a severe toll on our well-being, so there is growing evidence that unpolluted natural soundscapes have excellent restorative qualities. People are known to recover from injury or illness more quickly if they are exposed to the subtle soundscapes of wild nature. By polluting them with traffic noise across pretty much the whole of our countryside, we are robbing ourselves of curative environments, of opportunities to heal. It is rather ironic that recent research has shown that we find traffic noise more bearable, and presumably therefore less injurious, if it is shielded by the broadcasting of artificial birdsong through loudspeakers.

Quietening our countryside and restoring its natural soundscapes will bring benefits for people and nature in equal measure. But we will need to do much more than that, because noise is far from being the only pollutant of traffication that causes widespread damage to both our health and our wildlife.

CHAPTER 7

Emission Creep

I n the spring of 2020 the citizens of Jalandhar, in India's northern state of Punjab, saw something that amazed them, something that many of the city's younger inhabitants had never seen before in their lives: the Himalayas. For the first time in decades, the snow-capped peaks of the Dhauladhar range, well over 100 miles away to the east, could clearly be seen. Residents flocked onto social media to post photos snapped from their rooftops of the distant wall of rock and ice. One of the city's most famous sons, the cricketer Harbhajan Singh, tweeted 'Never seen Dhauladhar range from my home rooftop in Jalandhar ... never could imagine that's possible.' Others pointed out that when the mountains dissolved into evening darkness, an equally spectacular natural phenomenon could be seen simply by looking upwards, because the stars had reappeared in the night sky.

The reason for these wonderful revelations was, of course, the COVID-19 lockdown. All over the world, restrictions were placed on people's movements to control the spread of the disease and lessen its appalling toll. Traffic fell for the first time in generations and air quality improved almost immediately, most noticeably in cities. Long-hidden vistas sprang back into view. For many people this came as a wake-up call, the most graphic illustration possible of just how much air pollution we live with.

We drive immense distances over our planet's fragile surface each year, 15 trillion miles or more. The Earth's ancient orbit around its sun takes a year, but our motor vehicles cover the same distance every 20 minutes. What makes these unimaginably long journeys more remarkable still is that they are powered not by some futuristic fusion engine, but by igniting the remains of microscopic sea-life that lived and died when dinosaurs ruled the Earth. Most of our driving is still fuelled by petrochemicals distilled from crude oil, and most of our oil started life as plants, algae and plankton in the shallow seas of the Triassic, Jurassic and Cretaceous periods (collectively known as the Mesozoic). The period we live in now has been named by some the Anthropocene, or the Age of Mankind. But others, recognising the immense impact that our love of burning things has had on our planet, have given it a different name: the Pyrocene, the Age of Fire. Combustion is the driving force of traffication; and what we ignite, we inhale.

The cremation of Mesozoic sea-life in our engines releases gases and other by-products into the atmosphere through the exhaust pipe (or tailpipe, if you are North American). Carbon dioxide, water vapour and nitrogen make up most of these gaseous emissions by volume. None of them is directly toxic to people or wildlife, at least in the concentrations released from traffic, but carbon dioxide is the single biggest contributor to global climate change. Road transport accounts for 15 per cent of global carbon dioxide pollution and 10 per cent of total greenhouse gas emissions; its contribution to climate change is rising faster than that of any other sector.[34] Hiding among the carbon dioxide and the water vapour are smaller quantities of far more toxic gaseous emissions, such as carbon monoxide, volatile organic compounds such as benzene and formaldehyde and, perhaps most harmful of all in the quantities in which they are emitted, nitrogen oxides (NOx).[35]

If gas molecules released by combustion are the grains of sand on a beach, then particulates are the pebbles and the boulders. Particulate matter (PM) is an aerosol of microscopic droplets of liquid, solid particles and combinations of the two. Suspended (airborne) PM includes natural particles – dust, sea spray, fungal spores and pollen among others – but also tiny specks of metals, microplastics, organic compounds, droplets of acid, ammonium salts, soot and a wide range of other substances that owe their presence in the air to human activities.

The particulates emitted directly through the exhaust are mostly flecks of soot with other chemicals bound to their surface. Thanks to massive improvements in engine efficiency and post-combustion filtering, this source of traffic pollution is becoming less serious as older, more polluting vehicles are gradually replaced by far more efficient newer ones. Once on the road or suspended in the air, however, pollutants from the exhaust (primary pollutants) quickly evolve into a complex cocktail of new contaminants (secondary pollutants), both gaseous and particulate. Just as sand on the beach can be bound into nodules by tar, and pebbles can be eroded to sand, so there is a two-way flux between gases and particulates. Solar radiation drives the oxidation of exhaust gases to create particulate droplets of sulphuric or nitric acid, which in turn react with other particulates or hitch a ride on them. NOx react with volatile organic compounds to form ozone, a primary ingredient of smog and a major cause of ill health. Driven by the power of the Sun, these particles bind to and react with each other to create new and often more dangerous secondary pollutants. The smallest particles of all can bind to a range of organic carcinogens such as benzene and formaldehyde and carry them far from the road. Pollutants from industry, construction and other sources settle on roads and react with those from traffic in ways we are barely even beginning to understand. It is a sub-microscopic ecosystem of chemical food-chains every bit as complex and dynamic as a tropical rainforest. Perhaps recognising this analogy, scientists studying air pollution often refer to different types of contaminants as *species*.

From the perspective of human health, scientists are interested as much in the size of particulates as in their chemical composition, because this determines where in our bodies they eventually end up and what damage they do. In less polluted parts of the world, larger natural particulates prevail – pollen, sea spray, dust and so forth. Our bodies have lived with these for millions of years and we have evolved defence mechanisms to deal with them, from nasal hairs to bronchial mucus. But over heavily trafficated areas, the air contains high concentrations of very much smaller particles. These are far more dangerous because they can slip through our defences and end up lodged deep within our lungs. The smallest of them continue their invasion of our bodies by passing through the lungs and entering the bloodstream, giving them access to all parts of our bodies. They can even cross the blood–brain barrier and penetrate our central nervous system; particulate matter recovered from human brains is consistent in size and composition with that generated by road traffic.

Scientists have therefore developed a system to classify particulates by their size. Those small enough to be inhaled deep into our lungs are designated PM_{10}, the PM standing for particulate matter (or material) and the number indicating their size, a diameter of 10 micrometres or smaller. Despite being less than a fifth the thickness of a human hair, the larger PM_{10} particulates, those with a diameter between 2.5 and 10 micrometres (often termed $PM_{10-2.5}$), are termed coarse particles. Those with a diameter smaller than 2.5 micrometres get an additional designation, $PM_{2.5}$, and are called fine particles. Much of the $PM_{2.5}$ pollution in outdoor air, particularly in cities, is generated by road transport; despite being smaller than bacteria, it was these particles that were largely responsible for hiding the Himalayas and the stars from a generation of Jalandhar's citizens. Smallest of all are the ultra-fine particles, those with a diameter smaller than a tenth of a micrometre ($PM_{0.1}$), less than one hundredth the thickness of cling film.[36] Despite having negligible mass they are overwhelmingly the most numerous particulates generated by motor vehicles, and often the most dangerous.

The wheels of misfortune

Think of air pollution from road traffic and you will probably form a mental picture of black smoke belching from a vehicle's exhaust; indeed, this powerful image has been used on the front covers of several books and reports about air pollution. But new research suggests that a picture of a car wheel might be a more appropriate illustration of the problem because, just as with traffic noise, the main source of particulate pollution has now shifted from the engine to the point of contact between the tyres and the road. Improvements in engine technology and post-combustion filtering, driven by tight regulations, have greatly reduced particulate pollution from fuel combustion and most new vehicles now produce lower emissions than legal limits allow.[37] But this has done nothing to reduce what are now traffication's dominant emissions of these dangerous specks of matter, the wear and tear of tyres, brake pads and the surface of the road itself, collectively known as non-exhaust emissions (NEE). Cars have become heavier and faster, increasing all three non-exhaust sources of particulate matter. A car being driven carefully at modest speeds on correctly inflated tyres produces one and a half *trillion* ultrafine particulates through tyre wear every kilometre. If tyres are under-inflated, if the driver accelerates and brakes hard, if the road surface is rough or if the vehicle is overloaded, then non-exhaust emissions rise exponentially. Non-exhaust emissions now comprise well over half of the primary particulate emissions from vehicles, and this proportion is rising as exhaust emissions fall. A leading British laboratory for vehicle emission measurement called Emissions Analytics has recently gone much further, suggesting that NEE might now exceed exhaust emissions of particulates by a factor of well over a thousand. The parts of our road system that generate the highest concentrations of particulates are not necessarily those carrying the greatest number of vehicles, but those where vehicles accelerate and brake the most, such as road junctions.

In terms of numbers, the great majority of the individual particles produced by tyre wear fall into the PM_{10} size class, and so are small enough to become airborne and contribute to air pollution. However, they represent less than a tenth of the weight lost by tyres through wear and tear. The bulk is in the form of much larger particles formed from an amalgamation of tyre and road surface material. These usually have a sausage-like shape and roll like tiny black snowballs over the road surface, gathering a complex mix of finer particles and growing in size. The reason tyres are black is because of the addition of a substance called carbon black, a dark sooty material usually sourced from the burning of fossil fuels that is used to increase tyre strength and resistance to UV light. It is a major component of road dust and is not biodegradable; it is also quite possibly carcinogenic. Tyre and road wear particles also contain toxic chemicals such as 2-mercaptobenzothiazole, used in the vulcanisation of rubber and again listed as a potential human carcinogen, and flecks of zinc and other metals that can be highly toxic if inhaled.

Most of the components of tyres are synthetic polymers derived from petroleum: to quote from a report by Friends of the Earth, 'tyres are essentially yet more big chunks of plastic'. Tyre wear particles are therefore often considered to belong to a class of pollutants known as microplastics. Some estimates suggest that tyre wear might be responsible for as much as half of all the microplastic pollution in the terrestrial environment. A car tyre sheds around 1 kg of material on the road during its lifetime, and a truck tyre around 8 kg; this might not sound like much, but multiplied over tens of millions of vehicles it means that tyre wear produces over 60,000 tonnes of pollution annually in the UK. Globally, something like 6 million tonnes of tyre wear microplastics are generated by road traffic each year, enough to fill the world's largest container ship more than 30 times over.

Tyres are not the only contributors to non-exhaust particulate pollution. Most vehicles are slowed by the pressure of brake pads on the moving wheels. This causes abrasion to both surfaces and generates the release of a wide range of particulate pollutants. Brake pad wear releases significant quantities of particulate iron, copper, zirconium and antimony plus a whole range of organic matter, much of it in the dangerous $PM_{2.5}$ size class. The harder the brakes are applied, the greater the production of smaller, and hence more dangerous, particulates. We know very little about the scale of the problem, but it has been suggested that the weight of brake pad particulates shed each year is about a tenth that of tyre wear – another three colossal container ships full of highly toxic dust being dumped each year into the world's soils, rivers and oceans. To this deadly fleet can be added at least one more massive ship full of microparticles from the paint that is worn off the world's road surfaces every year.

Abrasion, and not combustion, is now the most important producer of particulate pollution from vehicles. As we drive along in our new electric cars, their batteries charged from renewable sources, we may feel smugly emission-free, but our tyres and brakes are spraying an invisible dust of toxic particles into the air that greatly exceeds the particulate pollution emitted from the exhausts of the combustion cars around us. And unlike exhaust emissions, which are tightly regulated and tested, there is nothing at all to limit the amount of non-exhaust emissions that cars can produce. In its 2019 report *Non-Exhaust Emissions from Road Traffic*, the UK Government's Air Quality Expert Group recommended that NEE are immediately recognised as a significant form of airborne particulate pollution, and that the problem applies equally to electric vehicles with zero exhaust emissions. But at the moment there is no incentive for tyre or brake pad manufacturers to make their products more resistant to wear and therefore longer lasting; indeed, they are not even obliged to reveal what compounds their products contain.

Pollution from exhausts and all the other toxin-shedding parts of motor vehicles is not the only entry in traffication's debit column when it comes to air pollution. Motor vehicles run on petroleum in two senses: the fuel that powers them and the asphalt that is used to make the smooth surfaces they require are just different distillations of the same crude oil, the petrochemical equivalents of skimmed milk and clotted cream. The web of hardened oil slicks in which we have willingly enmeshed ourselves is a significant source of pollutants in its own right, exuding complex mixtures of organic compounds that combine in sunlight to form hazardous particulates. The problem is particularly severe on hot days when these solidified slicks start to soften, sweating an invisible fog of toxic compounds that rolls over adjacent fields, gardens and playgrounds. At such times, the road surface can be a greater source of air pollution than the traffic it carries.

Unlucky number four

Air pollution, just like noise pollution, is a killer. The statistics are quite shocking. At least 4 million people, and perhaps as many as 9 million, die each year around the world from the effects of outdoor air pollution, with particulate matter being the biggest killer. One in eight deaths in the EU is attributable to air pollution. In 2019, 300,000 premature deaths across the EU were linked to chronic exposure to fine particulate matter ($PM_{2.5}$), 40,000 to long-term nitrogen dioxide exposure and 17,000 to acute ozone poisoning. A high proportion of these deaths can be linked directly to road traffic. In the UK, road traffic pollution is thought to account for over half of the 30,000 premature deaths caused each year by air pollution, and tyre wear particles alone may kill 8,000 people.

It is almost quicker to list the health problems that air pollution doesn't cause than to itemise those it does. A review of the scientific evidence published up to 2019 found that particulate air pollution can reach, and can affect, virtually every cell in the body. As well as diseases associated with the respiratory and circulatory systems – asthma, bronchitis, lung cancer, stroke, heart disease, intravascular coagulation, heart attack and a host of other killers – air pollution has been linked to depression and schizophrenia, all forms of dementia, pre-term birth, kidney disease, liver disease, leukaemia, osteoporosis, conjunctivitis, inflammatory bowel disease, skin disease and many other foul and life-eroding conditions. Falling birth rates in industrialised nations have been linked to increased exposure to the by-products of burning petrochemicals. Children exposed to air pollution suffer reduced growth rates, both physically and mentally. Sometimes it is hard to believe that we inflict this toll of suffering on ourselves more or less willingly.

Exposure to particulate pollution over the long term damages people's respiratory, cardiovascular and immune systems, greatly increasing their susceptibility to airborne infection. The newly clarified air of the COVID-19 lockdown did far more than give the residents of Jalandhar their first views of snow-capped mountains; it also directly reduced the spread and impacts of the disease. Lockdowns were imposed to curtail the movements of people in an attempt to slow transmission rates of the disease from person to person, but the incidental benefit of cleaner air might have contributed even more to containing the pandemic. Research from several countries has shown that lower levels of air pollution (particularly nitrogen dioxide) were associated with lower numbers of COVID-19 cases and lower death rates: if you lived downwind of a busy highway when the pandemic struck you were more likely to catch COVID-19, and more likely to die from it, than if you lived upwind.

Air pollution usually causes health problems that are gradual and long-term (or chronic), making it hard to establish a firm link between the problem and its cause, but it can also bring about immediate (acute) symptoms. Perhaps the most dramatic evidence for an acute impact of traffic pollution on human health comes from a study in China's crowded capital city Beijing, where car emissions have now overtaken the burning of coal as the leading source of air contamination. Resourceful researchers there hit upon an ingenious way of overcoming a lack of data on traffic flow or on individual health outcomes. The authorities in Beijing reduce congestion by banning cars from the road on certain days according to the last digit of their number plates; those ending in a 1 or a 6 are banned from using public roads one day each week, those ending in a 2 or a 7 are banned on another day, and so forth. When it comes to the turn of the number 4 to be banned, however, traffic levels are higher than on other days because 4 is an unlucky number in China and many people avoid number plates bearing that digit. The researchers found that on days when 4s are banned, congestion is higher than on other days, pollution levels rise and more calls are made to the Beijing Emergency Medical Center requesting ambulances. This is not because of an increase in the number of accidents, but to a spike in a range of acute heart and respiratory problems. The number 4 certainly proves to be unlucky for some.

Just like noise pollution, the burden of traffication is not spread evenly throughout society. People from the most affluent areas contribute disproportionately to the many problems of air pollution by driving further each year in larger vehicles, yet they tend to live further from busy roads. Households in the poorest areas emit the lowest per capita amounts of NOx and particulate matter, but suffer the highest levels of air pollution.

Pollution gradients

A devastating article published in 2021 by Ben Phillips and colleagues at the University of Exeter suggested that traffic-generated emissions pollute pretty much every part of Britain's land. Making informed assumptions about how far pollutants such as light and heavy metals, nitrogen oxides and particulates spread outwards from roads and into surrounding areas, and combining this with a map of Britain's roads (as we did with noise pollution in Chapter 6), they suggested that at least 70 per cent of our land, and perhaps over 90 per cent, is contaminated by pollution from motor vehicles. Just as with traffic noise, almost nowhere escapes.

But the same is true of air pollution from other sources, too. The gases and particulates we inhale with each breath, and which shower down constantly on our countryside, come from many different sources, both close at hand and far away. Road transport accounts for around a quarter of all the UK's PM_{10} emissions (but nearly 40 per cent of the more dangerous $PM_{2.5}$ component and perhaps over half of ultrafine $PM_{0.1}$). The rest comes from power plants, factories, forest fires, domestic wood-burners, lightning and a range of other sources; even trees can produce chemicals that react with NOx from car exhausts to produce secondary particulates. Smaller particles can be blown around in the atmosphere for days or weeks until rainfall brings them down to earth, making it very hard to trace them back to their sources. They may have their origins in other countries, perhaps even in other continents. The UK exports more public health damage to the rest of Europe than it imports, since the prevailing westerly winds carry the pollution of our cars eastwards over the continent, while blocking flow in the opposite direction. It has been estimated that $PM_{2.5}$ pollution produced in China is linked to around 65,000 premature deaths each year in other regions of the world, including over 3,000 deaths in the USA and Europe.

This presents road ecologists with a problem – how is it possible to isolate and study the impacts of road vehicle pollution on wildlife when it is so mixed up with pollutants from other sources? This is a problem that does not apply to most other areas of road ecology. Noise can easily be traced back to its source, allowing scientists to be sure that most of the noise we experience in our daily lives comes from road traffic. Studies of the movements of animals and their genetics clearly show that roads act as barriers to wildlife. Roadkill is a problem that can be even more unambiguously attributed to traffication – what else but vehicles could be flattening so many animals on the roads? But it is much more difficult to isolate the unique impacts on wildlife of air or water pollution from road traffic.

This might explain why road ecologists have largely shied away from studying the issue. An article published in 2017 (entitled 'Exhausting all avenues'!) recognised that while road ecology has grown into a dynamic multidisciplinary research area within the environmental sciences, the impacts on wildlife of chemical and particulate pollution from traffic have received relatively little attention.

But there is one consistent pattern that helps the few road ecologists working in this field of research. Just as traffic noise falls away with distance from a road (see Figure 5, p. 107), so too do the concentrations of most other pollutants from vehicles. Roadside soils are 'deeply enriched' (a technical but slightly euphemistic term meaning heavily polluted) by a range of heavy metals and semi-metals, many of them highly toxic. Lead, cadmium, copper, cobalt, zinc, chromium, arsenic, antimony and others all occur in high concentrations in roadside soils, but these elevated levels of contamination rarely extend much beyond 20 m from the verge.[38] Larger particulates, such as tyre and road wear particles, are generally deposited within a similarly narrow band along roadsides. For smaller particulates, elevated levels of contamination are often apparent 100 m or more away from the road, and for the lightest pollutants, such as ozone and NOx, the zone of detectable contamination may extend for 300 m or more, particularly downwind of the road.

This gives scientists a gradient of pollution to work with; they can study wildlife species or communities close to the road and compare them with those further away, and reasonably assume that any differences they find, all other things considered, are due to pollution that comes from that road. Because the impacts of air pollution manifest themselves in subtle ways that require road ecologists to examine their subjects closely, indeed often internally, and because researchers do not want their subjects messing up their results by moving between areas of low and high pollution, most of these roadside pollution studies have been undertaken on plants and invertebrates.

The results of these studies suggest that wildlife near roads faces a barrage of pressures from traffic pollution. An overview published in 2016 concluded that vegetation is significantly impacted by exposure to motor vehicle pollution, particularly NOx and ozone, both of which are toxic to many plants and their pollinating insects. Roadside trees show signs of poor health, a reduced density of photosynthetic pigments in the leaves, a shorter growing season and higher levels of damage by insect pests. The diversity of epiphytes such as mosses and lichens on trees falls as NOx levels rise. Root growth is suppressed by high levels of roadside lead and cadmium, and plants exposed to traffic exhaust fumes contain fragments of damaged chromosomes in their cells.

However, those plants capable of withstanding the toxic properties of NOx are able to benefit, because nitrogen is a fertiliser. Several species of common grasses are particularly adept at profiting from this fertilising smog of nitrogen and can grow in dense swards along polluted roadsides, to the exclusion of almost everything else. It is because of this fertilising effect that roadside nature reserves need frequent cutting to prevent flower-rich meadows from being choked by stands of rank grass. Of England's nitrogen-sensitive wildlife habitats, 95 per cent are thought to be adversely affected by nitrogen deposition, much of it from road traffic. These effects are strongest within 100 m of the road and are often detectable up to 200 m; downwind, the effects on roadside plants of toxic gases from traffic might be discernible for 300 m or more. If we look back again at Map 2 (p. 37), which shows all the parts of Britain that lie within 500 m of a road, we can get some idea of the area likely to be affected.

These patterns concern us all, and for more than just reasons of our health. The world's human population has passed the 8 billion mark and is speeding towards a predicted 10 billion by around the year 2060. New ways urgently need to be found to increase food production to meet this growth without adding to the existential problem of climate change. Air pollution creates a toxic haze that reduces the growth of crops by directly poisoning them and by reducing the solar radiation they receive to power their photosynthesis. Road dust settling on the leaves of plants, including crops, reduces their photosynthetic capacity still further. In China, the potential yields of rice and wheat are reduced by as much as a third in the worst affected places, resulting, in the case of wheat, in an 8 per cent reduction in yield across the entire country. Reducing NOx emissions might raise crop yields across the planet by as much as 10 per cent. We are not only poisoning ourselves with air pollution from road traffic; we also risk starving ourselves.

Plants are not the only organisms affected by traffic pollution. The delicate scent of flowers is produced to attract their insect pollinators, but NOx and other gases from vehicle engines chemically degrade these floral odours almost as soon as they are produced. Experiments have shown that honeybees accustomed to following these natural scents back to their pollen-rich source have a poor ability to detect chemically altered odours. Exhaust fumes might also impair the bees' cognitive ability to recognise and remember plant odours. Clever experiments carried out by researchers at the University of Reading in southern England found that artificially raised levels of diesel exhaust fumes and ozone in open fields led to a fall in the number of insect pollinators of up to 70 per cent, and a drop in the number of visits made to flowers of up to 90 per cent. What is so alarming about the results of this study is that the elevated levels of NOx and ozone that caused these huge declines in pollination services were significantly lower than the limits considered safe for people under current air quality standards.

Pollinating insects also risk absorbing particulate traffic pollution when visiting roadside flowers and carrying pollen back to their nests. Research in Italy has shown that when honeybees fly from flower to flower collecting pollen and nectar, the fine hairs that cover their bodies generate an electrical charge. When bees forage close to roads, this weak static charge causes ultrafine particulates from tyre and road wear, particularly barytes and various oxides and hydroxides of iron, to stick to their bodies and to the pollen they collect. This contamination is then taken back to the hive and ends up in our honey. Foraging bees may also accumulate heavy metals such as lead, chromium and cadmium as they visit contaminated roadside flowers. Other invertebrates such as roadside snails can similarly accumulate high concentrations of lead, cadmium and zinc in their bodies.

How the airborne pollutants of traffication affect larger and more mobile species is unclear. Birds have evolved particularly efficient respiratory systems to allow them to meet the demands of flight; these function in a different way from those of mammals and might be more vulnerable to gaseous and particulate pollution. What little we know suggests that, just like ourselves, birds suffer respiratory distress, higher risk of infection and a range of other symptoms when exposed to the air pollution of traffication. A remarkable study from the USA has suggested that the impacts of air pollution on birds might be substantial and perhaps devastating. Using data collected by thousands of amateur birdwatchers, the researchers found that areas with higher ozone concentrations have suffered significantly greater losses of birds, and that measures introduced to cap ozone emissions from industrial sources to safeguard human health may also have prevented the further loss of more than a billion birds. Traffic pollution is a significant and pervasive source of ozone, and the results of the US study suggest that air pollution from cars has the potential to drive down bird populations across vast areas. The study was unable to identify the causal mechanism behind this pattern, but the researchers speculated that it might result from the poisoning effects of ozone on insect populations, leading to reductions in the numbers of birds that rely on insects for food. Some evidence to support this theory comes from a study of house sparrows in the UK, which suggested that air pollution from traffic can reduce the availability of insect food for chicks, causing them to starve.

The salmon's warning

Microplastic pollution from tyre and road wear is washed from roads by rain or blown by wind into adjacent soils, where much of it becomes trapped, but some of it is carried by surface water into streams and rivers and thence to the sea. Tyre wear may be responsible for up to 10 per cent of all the microplastic pollution entering the world's oceans (some estimates suggest it may be closer to 30 per cent, and one has even put it as high as 60 per cent). Because the specific gravity of tyre material is usually slightly higher than that of seawater, it sinks very slowly to the seabed, often after being carried far out to sea in ocean currents. In deep ocean trenches yet to be explored, our pollution precedes us in the form of a steady rain of microplastic particles from tyres being worn down on roads thousands of miles away. Microplastic particles collected from Arctic sea ice and Alpine snow are consistent in their chemistry with those shed by car tyres.

Tracing these tiny particles back to their sources is extremely difficult and requires detailed chemical analysis, but more difficult still is working out what impacts they have on the environment. Microplastics from sources other than tyre wear have been found to harm marine worms by blocking their guts and by acting as Trojan horses, smuggling toxic chemicals into their digestive tracts. In soils, microplastics can have profound impacts on the composition and abundance of invertebrate communities. Until recently, however, tyre wear particles were thought to be relatively benign to wildlife. Researchers who exposed three freshwater species (an alga, an invertebrate and a fish) to tyre wear particles found that it was not toxic to them at levels likely to be found in the environment. A number of subsequent studies also proposed that tyre wear is unlikely to be a serious pollutant in aquatic ecosystems. But a recent case has undermined this complacency, and pushed tyre wear particles right to the forefront of water pollution research.

For many years, scientists were puzzled by the mass deaths of a fish called the coho salmon in the waters of the Pacific North-West of America. Salmon returning from the sea to the streams and rivers of the Puget Sound, hoping to end their days in a final frenzy of spawning, have been thwarted in their lusty ambitions by dying in huge numbers before reaching the spawning grounds. Half of the coho salmon that return to Puget Sound's urban streams die before they can spawn, and in some streams *all* of them die. The fish show symptoms characteristic of poisoning, appearing to gasp for breath and swimming aimlessly around in circles before dying. Juvenile salmon migrating in the opposite direction also die in large numbers, and with the same symptoms. The Puget Sound is ringed by the dense human population of Seattle and its satellite cities, and scientists studying the problem noticed that the mass die-offs of coho salmon occurred shortly after periods of heavy rain, so it occurred to them that perhaps runoff was washing pollutants into streams from surrounding urban areas.

It was not until December 2020 that they uncovered the source of the problem. All the water samples they collected contained tyre wear particles, and the scientists found that when these particles – and nothing else – were added to water, they induced the same symptoms in coho salmon as those seen in fish dying in shoals in the streams of Puget Sound. Having identified tyre wear as the source of whatever was killing the salmon, the scientists then faced the significant problem of identifying which of the many hundreds of different chemicals they found in their tyre wear solution was responsible. By painstaking analysis, testing compound after compound, they eventually alighted on one that caused salmon in testing tanks to exhibit exactly the same symptoms as those shown by wild fish. The culprit turned out to be a chemical called 6PPD-quinone (6PPD-Q).[39] The reason it took so long for the scientists to identify this chemical was that until it was detected by their investigation into the coho salmon die-offs, nobody knew that 6PPD-Q even existed. It turns out that it is formed by a reaction between 6PPD, a chemical used in tyres as a preservative, and ozone (itself a pollutant created by road traffic). Although it differs from 6PPD by just a couple of oxygen molecules, 6PPD-Q has very different properties to its parent; it is far more soluble in water and it breaks down in the environment much less quickly. And as far as coho salmon are concerned, it is a hundred times more toxic.

Subsequent tests have found that two species of trout, fish that are closely related to salmon, are also highly susceptible to 6PPD-Q poisoning.

However, other fish such as char and sturgeon, and aquatic invertebrates such as crustaceans, appear to be far more tolerant of the chemical. Remarkably, even some other species of salmon appear to be unaffected; chum salmon show no symptoms of poisoning when swimming in water containing concentrations of 6PPD-Q that quickly kill coho salmon.

The case of the coho salmon provides a stark warning. If the coho salmon had not been a commercially and culturally important species, or had it happened to be one of the fish that is not susceptible to 6PPD-Q, then the investigations would never have taken place and the existence of an acutely toxic pollutant of tyre wear would still be unknown. Now that it has been identified, 6PPD-Q is turning up everywhere; it has been found in water, air and soils from Toronto to Hong Kong, and Australian scientists have found it, in concentrations similar to those killing Seattle's salmon, in a creek near Brisbane. But the problem does not stop there; once 6PPD-Q was identified, chemists looked to see what other compounds might be formed by the reaction of 6PPD with ozone – and discovered no fewer than 38 of them! We know nothing about the toxicity of all these newly recognised pollutants to wildlife. Indeed, we know nothing about their toxicity to ourselves.

As 6PPD is used by practically all the world's tyre manufacturers, future work will no doubt show 6PPD-Q to occur wherever there are roads – which in many countries is more or less everywhere. Unfortunately, there are no known alternatives to 6PPD that provide the same safety and performance characteristics in tyres.

Nobody knows why coho salmon are so extremely sensitive to 6PPD-Q while other fish, even other salmon, appear to be unaffected by it, but examples of highly selective toxicity like this are not uncommon in nature. The catastrophic collapse of India's vulture populations has been caused by a veterinary drug given to cattle called diclofenac. Vultures are acutely sensitive to diclofenac and die within a few hours of picking meat from cattle carcasses that contain even the faintest traces of the drug, yet many other species of birds are quite unaffected by it. Closer to home, pet owners know that you should never feed chocolate to dogs, or let cats ingest the pollen from iris flowers, or allow rabbits to eat avocado. But while this kind of selective toxicity is fairly common in nature, it is utterly unpredictable. There is no way of even guessing which of the thousands of different chemical pollutants produced by traffication – and the tens of thousands of secondary compounds they form by reacting with other pollutants in the environment – will prove to be toxic to which of the tens of millions of animal and plant species on the planet.

The coho salmon story raises the question of how many other unknown toxins from tyre, brake and road wear are leaching into our rivers and oceans, and what impacts they may be having there. Research showing that pollutants are safe when tested on a handful of species in a laboratory might tell us nothing about how they affect other species in the wild. There is simply no way of knowing how much this rich and highly dynamic cocktail of contaminants has contributed to the loss of our wildlife; we have barely started to look. As scientists like to say, absence of evidence is not the same as evidence of absence. Or, in plain terms, just because we are not aware of such problems does not mean that they aren't going on all around us.

The salt road

One chemical pollutant that can be unambiguously blamed on traffication is salt. Rates of vehicle ownership are often high in temperate regions, because that is where the planet's wealth tends to be concentrated. It is also where most of the planet's freshwater lakes are found. What this means is that people in cold countries tend to do a lot of driving, much of it near open water. In the USA, 70 per cent of the human population lives in regions that regularly experience snow and ice, and in the UK it is almost all of us. When temperatures drop below zero and ice starts to form, the gritter trucks take to the roads to help keep the traffic moving and reduce the number of accidents.

Road salt, often called grit but usually the same sodium chloride that we sprinkle on our food, is a serious pollutant with significant impacts on biodiversity.[40] Around 60 million tons of salt, a fifth of the planet's total production, are dumped on the world's roads each year, 2 million of them in Britain. Despite ever-milder winters as climate warming starts to take hold, the amount of salt being spread on the world's roads each year is actually increasing, in part because China has recently become an enthusiastic user. Rain, spray from vehicles and meltwater cause the dissolved salt to run off roads and into adjacent watercourses and ponds, changing their chemistry and their electrical conductivity with significant impacts for aquatic life. In extreme cases, ponds and streams close to roads have concentrations of chloride one hundred times higher than unaffected watercourses, a salinity almost a quarter that of seawater. Road salts can spread through groundwater flow and storm drains, contaminating wetlands up to 200 m away. This pollution does not end with winter's last frost or its final dump of snow, because salt is a persistent pollutant. A recent study in Canada found that even in the height of summer, nearly 90 per cent of waterbodies near roads have chloride levels that exceed federal guidelines.

The environmental impact of salination in freshwater ecosystems has proved a particularly fertile field of research for scientists in the USA, where up to 20 million tons of salt are spread on the highways each year. Thousands of the country's lakes are now known to suffer dangerously high levels of salination. This environmental threat, only recently recognised, has sparked a flush of new research that is yielding some unsettling results. Frogs living near salted roads suffer from oedema, a bloating of the body caused by fluid retention, which restricts their mobility and reduces their lifespans. Similar effects have been observed in tadpoles exposed to road salt. At higher concentrations, salt can alter the behaviour and growth rates of both tadpoles and adults and can kill them outright, but some species are more susceptible than others, resulting in changes in the entire amphibian community. Salt also impacts frogs' ability to survive the winter by reducing their natural anti-freeze, it makes them more vulnerable to diseases, it affects their immune systems and, in larval stages, it interferes with the functioning of the gills and leads to a higher rate of physical defects during development.[41]

Intriguingly, there is some evidence that invasive amphibians such as the African clawed frog, introduced by accident to North America and now swarming over the continent, are better able to cope with road salt pollution than native species. Just like the cane toad in Australia, invasive amphibians seem to have the edge over native ones in coping with road traffic – or, put more logically, their ability to cope with traffication helps to *make* them effective invaders.

The complexity of some of the relationships between road salt pollution and amphibian biology that have recently been uncovered suggests we are still a long way from understanding the extent of the problem. For instance, road salt swings the sex ratio of frog tadpoles towards males, but this depends on the type of leaf litter at the bottom of the pond. Oak leaves swing tadpole sex ratios towards females, so the addition of salt counteracts that impact. Maple leaves have no such effect, but a combination of salt pollution and maple leaves enhances the differences in size between male and female tadpoles, producing larger females. Another study has shown that low concentrations of road salt may actually have a positive impact on the growth rates of tadpoles, which grow more quickly and reach larger size, but this benefit is negated when they metamorphose into frogs, since those raised in saline water die sooner than those that developed in unsalted water. How all these complex effects play out in terms of driving changes in populations is unknown, but there is a growing recognition that road salt may prove to be a very important factor in the global collapse in amphibian populations.

The effects of road salt on smaller forms of freshwater life are still less well understood, but there is ample evidence that entire aquatic ecosystems can become unbalanced and food webs radically altered. Road salts can cause blooms of algae, utterly changing the structure of freshwater communities, and the diversity of microscopic life declines as species able to tolerate salt proliferate and the more sensitive ones disappear. The entire functioning of waterbodies can change. Saline water is heavier than fresh water and so it sinks to the bottom of lakes, forming a denser layer. This leads to a phenomenon called meromixis, the complete and permanent separation of different layers of water, with no mixing of the deeper layers with those from the surface. The lower layers of meromictic lakes are therefore virtually devoid of oxygen. This oxygen depletion releases phosphorus from lake-bottom sediments, killing off all the remaining life that requires oxygen and allowing anaerobic life to proliferate. Additional problems then arise, because anaerobic respiration in wetlands is the largest natural producer of methane, one of the most potent greenhouse gases. This link between road salt and climate change has only recently been discovered, and there are doubtless many further revelations to come about the environmental costs we pay for keeping the roads open when the thermometer drops below zero.

Plants exposed to salt spray near the roadside suffer reduced growth and poorer health, making them more vulnerable to infestations by pests such as aphids. Plant roots, including those of commercially important species such as onions, suffer reduced growth with rising levels of salt. Birds, too, may be vulnerable to the impacts of road salt, particularly where it is spread in larger granules. Large numbers of siskins and waxwings have been found dead on roads after eating salt crystals, perhaps mistaking them for grit. Some animals actually like road salt, as it provides them with a handy source of sodium. Studies of moose fitted with tracking devices have shown that normally the animals avoid roads as much as possible, rarely venturing within half a kilometre if they can help it. But the allure of the salt in roadside pools and vegetation sometimes proves too much for them to resist so they move towards danger. Many are killed on the roads as they try to exploit this saline harvest.

As always, there is a human health dimension to this pollution. Drinking water supplies can become contaminated by salt. Road salts can cause the release of highly toxic heavy metals such as mercury and lead from contaminated sediments and can erode metal pipes carrying drinking water, releasing other poisons such as zinc. Mosquitoes, a pest at best and a vector of deadly disease at worst, appear to benefit from saline water, perhaps because their aquatic larvae face less competition for food and space from other less salt-tolerant invertebrates.

An unhealthy glow

As darkness falls, diurnal animals retreat into the shadows and the bigger-eyed, larger-eared night shift comes on duty. Around a third of all the world's vertebrate species, and nearly two-thirds of invertebrates, are nocturnal. The abrupt change in natural illumination that occurs twice each day comes as a surprise to neither group of animals, because this cycle of light and darkness has been the only unchanging feature of Earth's environment since the first signs of life appeared in its oceans 4 billion years ago. Temperatures have soared and plummeted, the composition of gases in the atmosphere and the acidity of the oceans have fluctuated wildly and continental plates have scudded to and fro across the face of the planet, splitting and colliding along the way. But all of life has evolved around an unchanging celestial regime of light and dark as precise and dependable as a Swiss watch. It is not a simple cycle; the daily rotation of the Earth on its axis causes periods of day and night, but a tilt on that axis means that the ratio of the two changes through the seasons, more so in higher latitudes. Added to this is the waxing and waning of the Moon, which determines how much sunlight is reflected our way at night. But it is a cycle that has been repeated over and over again since life began. Life has evolved to depend on this one constancy in a chaotically changing world, and the light–dark cycle regulates everything from the activity of individual cells in our bodies to the functioning of entire ecosystems: it is the metronome that sets the rhythm of nature. The vertical migration of plankton and their predators in the oceans, the timing and direction of migration in birds, the spawning of crabs and frogs and the flowering seasons of plants all depend on a repeating barcode of light and dark that has not changed since life began.[42] Plants are so finely attuned to this cycle of light that in the tropics, where there is little difference between the year's longest and shortest days, some trees are able to detect changes in day-length of as little as 30 minutes and use this as their cue for flowering.

But this ancient cycle of light and dark has been disrupted. As the Sun goes down on our heavily trafficated world, lights flicker on along hundreds of thousands of miles of highway, and millions of drivers start reaching for their headlight switches. Previously dark areas are now illuminated, and by light in quite different wavelengths to that cast by the Moon and the stars.

Traffication is far from being the only source of artificial light at night (or ALAN, as scientists like to call it). Light from homes and offices, factories, stadium floodlights, airport runways, advertising hoardings and other sources adds to that from cars and streetlights, and all of it is scattered in the atmosphere by water droplets and other airborne particles (many of them also the pollutants of traffication) to create a phenomenon called skyglow. This is clearly visible from afar as a dome of luminescence that sits over cities and towns at night, the glowing nocturnal equivalent of the grey haze of air pollution that covers them by day. In heavily trafficated countries, skyglow pollutes night skies almost everywhere; it can be bright enough to cast shadows of trees in woodland several kilometres from the sources of light that create it, and the glow of large cities may be visible in the night sky for hundreds of kilometres. It completely reverses natural patterns of nocturnal illumination; skyglow is brightest on cloudy nights, when natural light would normally be at its faintest. Perhaps a quarter of the planet's surface is polluted by artificial light at night, but in densely populated and heavily trafficated countries such as the UK almost everywhere is affected. Less than a tenth of the population of the UK, and only a third of the world population, can now see the Milky Way from where they live. Natural darkness, illuminated only by the Moon and stars, has become so threatened in many countries that some of the remaining areas are now protected as dark sky parks and reserves. Around half of all the world's Key Biodiversity Areas, a network of sites of supreme global importance for the preservation of life on Earth, lie entirely under artificially bright skies, and less than a third retain all their natural darkness.

Because light pollution has many different sources, each producing different spectra and all of them mixed in the scatter that produces skyglow, traffication's contribution to the problem is almost impossible to assess. However, it is likely to be both significant and growing. Roads permeate the landscape like no other light source, and few places are free of light pollution from car headlights and street lighting.[43] Professor Kevin Gaston of the University of Exeter, a world expert on the environmental impacts of light pollution, has argued that 'emissions from vehicle headlights need to be considered as a major, and growing, source of ecological impacts of artificial night-time lighting'. Not only do vehicle headlights focus light to a greater intensity than almost any other source, they also project it in a horizontal plane, meaning that it can be visible to terrestrial animals over long distances. A car with its headlights on full beam transforms a dark night into an overcast day. It may take half an hour or more for the sensitive eyes of nocturnal animals to recover after being dazzled by headlights. Owing to the movement of vehicles, animals near the road experience headlights as pulses of light, which causes more disturbance than a continuous beam (indeed, flashing lights have been deployed deliberately to disturb animals near roads and drive them away).

Many car headlights, and increasing numbers of streetlights, now use light emitting diodes (LEDs) that emit broad-spectrum white light. This is rich in short-wavelength blue light known to interfere with many biological processes, including sleep. LED streetlamps are cheap, energy efficient (allowing them to be self-charged by small solar panels) and produce a white light that many drivers find preferable to the old orange sodium glow, but by recreating daylight at night they disrupt a wide range of biological processes and affect a large number of species.

Light pollution upsets everything. Diurnal species, even those that need enough light to see colour to find food, are now able to invade the night, bringing unwanted competition for nocturnal species. Trees near street-lights shed their leaves later in the autumn or not at all, providing a surface for fungal disease to enter, and they come into bud earlier in the spring, making them more vulnerable to late frosts. Bats that are happy to cross unlit roads will often avoid lit ones, particularly those lit by LED, causing their landscapes to become fragmented. Animals using the stars to navigate can become disoriented by light pollution from cities a hundred miles away or more owing to the long reach of skyglow. Hatchling sea turtles, programmed by evolution to move towards the reflected light of the Moon on the sea, become disoriented by artificial lights and move instead in the opposite direction, with usually fatal consequences. Female glow-worms hoping to attract a mate with the light of their luminous abdomen are outcompeted by light pollution and die without reproducing.

Entire communities and food chains can be disrupted, and they are disrupted in different ways by light of different wavelengths. Light pollution affects the hunting rates of predators and the avoidance behaviour of their prey. Aphids gather at night in greater numbers on more brightly lit plants because the parasitic wasps that attack them hunt less efficiently there. There is growing evidence that the global collapse in insect populations, one of the most worrying environmental crises facing mankind after climate change, has been caused at least partly, and perhaps largely, by light pollution. One of the most rapidly declining groups of insects, the moths, appears to be particularly susceptible. Moths may become disoriented by light from streetlamps hundreds of metres away. Street lighting reduces moth caterpillar abundance by up to half, disrupts their feeding and alters their rates of development. Adult moths in artificially lit areas struggle to find a mate and many die without breeding. Many plants rely on night-flying insects such as moths for their pollination, but this vital ecosystem process becomes disrupted in the presence of artificial light.

Light pollution is perhaps the least understood of all traffication's many contaminants, but because the planetary clockwork of light and dark is intrinsic to so many different natural processes, and because life on Earth has no evolutionary experience of anything else, it may prove to be one of the most devastating.

We are not immune to the destabilising effects of light pollution. Our body clocks, like those of most other animals, are set by the natural cycle of light and dark, and their disruption by light pollution has been implicated in a range of health problems, including sleep disruption, hormone imbalance, cardiovascular disease and cancer. There may even be a link between street lighting and diseases that have leapt from animals to people (zoonoses); bats have been implicated as the source of a number of zoonoses, including COVID-19. Road lighting causes insects to congregate around streetlights, attracting bats and bringing them into closer contact with people. Noise and air pollution from road traffic reduce the bats' immune systems, causing them to carry, and therefore to shed, higher viral loads. Furthermore, pollution from traffic impairs our own immune systems too, perhaps making it more likely we catch and transmit disease.

Eventually the eastern sky starts to brighten, and lights are switched off. Nocturnal animals go into hiding and diurnal life begins to assert its claim on the new day. The dawn chorus starts up, birds shouting threats and messages of love to each other over the growing rumble of early rush-hour traffic.

Our mechanised assault on the natural world begins again.

CHAPTER 8

In the Zone

Perhaps the single biggest issue in British wildlife conservation over the last 30 years has been the collapse of biodiversity on our farmland. Rachel Carson's famed book *Silent Spring,* published in 1962, evoked the spectre of a countryside robbed of its birdsong by the insidious impacts of farming with chemicals. But by 1990 the toxic pesticides at the heart of her bleak vision had largely been replaced by safer products.[44] That year, however, saw the publication of another book that, while never achieving anything like the fame of *Silent Spring,* was equally influential in British conservation. *Population Trends in British Breeding Birds* (1990), produced and published by the BTO, is an outwardly unassuming little book with soft blue covers and a rather homespun feel. It is seldom referred to these days: long out of date and long out of print, its survival on the competitive cliff ledge of the ornithologist's bookshelf relies increasingly on its historical significance.

But at the time it was transformative, because its central message was that many birds previously assumed to be too common and widespread ever to require the attention of conservationists were actually in severe and protracted decline. Species such as the corn bunting and the skylark, formerly the epitome of abundance and plenty in rural habitats, suddenly emerged as high conservation priorities. It soon became clear that these declines were indicators of a general collapse in the ecological health of the farmed landscape; populations of mammals, insects and wild plants have also suffered huge losses. Almost three-quarters of our land is used for food production of one kind or another, so these losses, which have yet to be reversed, represent a severe impoverishment of the country's natural resources.

The British silent spring arrived by stealth, and it coincided exactly with a period of unprecedented change in farming practices, fuelled after 1973 by the yield-boosting subsidies of the Common Agricultural Policy. It seemed certain that the loss of farmland wildlife had been caused by these profound changes in farming methods – but which ones? The catch-all term *agricultural intensification* covers a wide range of changes: different crops, different ways of growing, protecting and harvesting them, and different ways of arranging them in the landscape. New research was required to understand the problem and work out how best to tackle it, and scientists – me included – poured out into the fields to try to find out what was going on.

During the 1990s, almost every farmland bird species had its own little group of dedicated researchers trooping around after it, searching for answers. I led a team looking at the problems facing skylarks; colleagues took on other declining farmland birds: linnets, yellowhammers, lapwings, corncrakes, turtle doves, corn buntings, tree sparrows, grey partridges, yellow wagtails and more. Other ecologists joined the ornithologists, studying the mammals, plants and insects whose populations on farmland had similarly dwindled. Our ranks were swelled by scientists from all over Europe, because the more we looked at the continent's farmed landscapes, the clearer it became that this was a pandemic problem. It was surely one of the biggest research programmes ever undertaken in the cause of European wildlife conservation. What started as a thin trickle of supposition quickly became a flood of scientific articles, books and conferences. And when all the evidence was in, the verdict of the scientists was unanimous: agricultural intensification is to blame.

To be fair to my fellow researchers, the results of all this hard work were rather more nuanced than that. The great insight of this huge body of research was that, in ecological terms, agricultural intensification is not one problem but many. Some species were found to have suffered from the increased use of agrochemicals, others from subtle changes in the timing of crop sowing and harvesting, or changes in hedgerow structure, or patterns of grassland cutting, or declines in summer or winter food, and so forth. Only by identifying all these different components of change, and the impacts that each has had on wild species, was it possible to develop practical solutions.

But the underlying message from scientists was the same – the modernisation of our farmland has been responsible for the loss of much of our countryside's wildlife. Solve this problem and our lost biodiversity will surely return. For me and my fellow researchers it seemed that the job was done: the killer of our countryside had been unmasked, put on trial and found guilty. The scientific case was closed.

But now, looking back, I think we missed something.

The hidden accomplice

A third of a century has passed since the publication of *Population Trends*, but the issue of biodiversity loss on farmland is still firmly on the conservation agenda, because while their declines have slowed, the populations of most species are still far from recovery. Smartly tailored solutions developed to allow wildlife and modern farming to co-exist often work very well, but they have not restored wildlife populations to what they once were. The BTO's Farmland Bird Indicator, an index that tracks the changing populations of 19 bird species of agricultural habitats, has levelled out in recent years but remains firmly stuck at below half its 1970 value. And now other groups of formerly common birds have joined the larks and the farmland buntings in vying for urgent attention.

The 2020 edition of the *State of the UK's Birds* report, the definitive annual publication by a consortium of the country's leading conservation and research agencies, showed that there are now almost 20 million fewer pairs of breeding birds in the UK than there were in the late 1960s. That's an average loss of 80 pairs of birds in every square kilometre of land. Britain is not alone in suffering such losses – some 600 million birds have been lost across the EU since 1980. Although the overall number of birds has stabilised in recent years, our avifauna is becoming less and less diverse; our seven most abundant species now account for an astonishing *half* of all the individual birds in the UK.

Analyses updating those first presented in the BTO's little blue book back in 1990 now add woodland specialists and long-distance migrants (species that breed here and spend the winter in Africa) to the groups of birds showing severe and sustained declines. Once again, these declines have coincided with environmental changes that would appear to explain them. The structure of Britain's woodlands has changed through overgrazing by deer and the abandonment of traditional management practices such as coppicing. The Sahel region of Africa, a vital staging and wintering area for our migratory birds, has suffered increasing rates of deforestation, habitat loss, drought and fire over the same period. Once again the scientists are out in force, not unreasonably focusing their enquiries on woodland management and on changes taking place along African migration routes, for here, surely, are to be found the root causes of the problem.

And once again, I think we are all overlooking something.

I am not questioning the immense body of research that proves clear links between agricultural intensification and collapsing populations of birds such as the skylark, turtle dove and linnet. This work has shown beyond any doubt that changes in farming practices have contributed to declines in our countryside's wildlife. The best proof of this is that when agriculture is made less intensive, for example where a farmer decides to turn to organic production, wildlife usually recovers to some extent. But for no species can

we prove that the modernisation of agriculture has been the *only* factor in
its decline, or indeed even the most important. There is nothing in those
hundreds of scientific studies that precludes the existence of a second,
perhaps even greater, cause of our country's ecological impoverishment.

It was recognised over a century ago that scientists often become so
besotted with their favoured theory that they try, usually (I hope) subconsciously, to bend all their evidence to fit that single hypothesis, blinding
them to the possibility of alternative, or additional, explanations.

Now the evidence is mounting that we called off the hunt for the killer of
our wildlife too soon, because it seems increasingly likely that our primary
suspect, agricultural intensification, has had an equally culpable accomplice.

I hope that the preceding chapters have persuaded you that roads
and their traffic represent a serious hazard, or rather a number of serious
hazards, to wildlife. Cars and other vehicles squash, scare, stress, suffocate
and confuse our animals and plants. Roads break our countryside down
into ever smaller fragments and prevent animals from moving between
them: our once continuous, interconnected landscape has been reduced
to a patchwork of little tarmac-walled prisons. Traffic affects how animals
behave, where they go and how long they live. Road pollution, whether
from noise or fumes or salt or light, causes birds to age prematurely and
sing differently, frogs to bloat or change colour, bats and owls to miss their
prey, and bees to become electrical blobs of flying contamination. The
functioning of entire ecosystems is affected through the interruption of vital
processes such as communication, gene flow, pollination, seed dispersal and
water oxygenation. These effects can be detectable over hundreds of metres,
sometimes a kilometre or more, from the highway itself – think back to the
distances that soundscape pollution can carry from major roads, and how
sensitive some species are to even low levels of traffic noise. We are only just
beginning to recognise some of the dangers; particulate pollution from tyre
and road wear may, with further research, prove to be a hugely significant
problem for freshwater and marine biodiversity, as the case of the coho
salmon showed. Ozone pollution, largely from car engines, might prove to
be one of the most serious environmental problems we have ever created
because it may be driving the collapse of our insect populations.

But none of this *proves* that traffication has directly and significantly
contributed to the drastic loss of our wildlife that has occurred over the
last half-century. It is not impossible that wild animals and plants are
somehow able to withstand all the problems of roadkill, fragmentation and
pollution that roads throw at them, and that while some individuals might
suffer, their overall populations may be resilient. Perhaps there are benefits

to living near roads that outweigh the heightened risks of collision, poor health or interrupted communication. The number of animals flattened on the roads each year, although staggeringly high, might be fairly insignificant compared with the numbers starving to death, dying of disease or being killed by predators.

Furthermore, only about 1 per cent of Britain's surface area is covered by roads, around the same as our lakes and rivers. Our National Parks cover more than ten times this area, and you can add to that the hundreds of nature reserves with which we are blessed. Woodland covers more than 13 per cent of the UK's land surface. The area of land that is managed partly or wholly in the interests of wildlife greatly exceeds that which has been tarmacked over for cars. Can something that occupies such a tiny fraction of our land surface, no matter how damaging, really drag down populations of wild plants and animals across the entire country? Is it even vaguely plausible that a growth in traffic on roads, which cover just 1 per cent of our land area, has had an impact on wildlife that is comparable to that of the intensification of farmland, which covers over 70 per cent?

If we want to know whether traffication has contributed, directly and significantly, to the loss of our biodiversity, we need more than the evidence of the previous chapters. To be able to point the finger of blame firmly at the car, we need proof that roads actually drive down wildlife populations, and over very large areas.

The search for evidence

A huge amount of research has been invested in understanding what lies behind the decline of our wildlife populations, but it has proved very difficult to link any species' falling numbers unequivocally to a particular environmental change. This is partly because declines in wildlife and changes in their potential causes have happened almost everywhere: we cannot compare wildlife populations in modern agricultural landscapes with those in areas that are similar in terms of climate, soil type and terrain but which continue to farm exactly the way they did in the 1960s, for the simple reason that there aren't any. The same is true of roads; almost nowhere has escaped the rise of traffication. A recent study of the impacts of roads on Europe's mammals concluded that 'the lack of areas that could be used as controls implies that scientists may no longer be able to measure the magnitude of road effects'.[45]

A second difficulty is that our wildlife has slowly ebbed away, 1 or 2 per cent each year, over half a century, over which time all the possible causes of those declines have increased equally gradually. Rarely is it the case that a species' population has crashed catastrophically in the course of a year or two and that a particular environmental pressure increased massively over the same short period, something that would make it much easier to link the decline to its cause.[46]

Ecology is a messy field of research when it comes to hard numbers; the natural world is such a phenomenally complex and difficult-to-measure place that nothing corresponds perfectly with anything else. Everything we count or measure is so intricately intertwined with, and dependent on, so many other things that when ecologists plot one set of measurements against another on a graph, they usually see an inconclusive scatter of points. But just occasionally, ecology throws us something like a straight line.

Figure 6 plots an index of the population size of British farmland bird species (the BTO's Farmland Bird Indicator, set to an arbitrary value of 1 in the year 1970) against a measure of agricultural intensity, measured here as average UK cereal yield. Yield (the amount of produce harvested per unit area, usually expressed in tonnes per hectare) is a good measure of agricultural intensity because it captures in a single number all the many measures that are taken to increase productivity – boosting yield is, after all, what intensification is all about. Each dot on the graph represents a different year between 1970 and 2020. The relationship between the two is remarkably tight; it is the sort of graph that physicists or chemists might take for granted

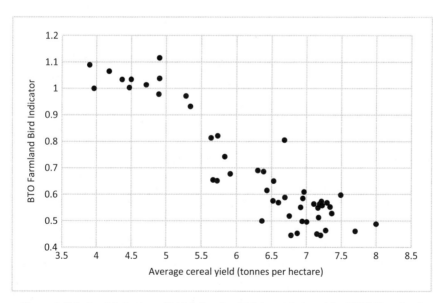

Figure 6. Relationship between British farmland bird populations (the BTO's Farmland Bird Indicator, set to an arbitrary value of 1 in 1970) and average cereal yield across Britain. Each dot on the graph represents a different year between 1970 and 2020.

but ecologists rarely see. Without exception, in years of low cereal yield the UK's population of farmland birds was high, and in years of high cereal yield it was low. A doubling of cereal yield over the past half-century has coincided with a halving of farmland bird populations. This strikingly clear pattern appears to confirm what we knew already from years of intensive field research: the loss of biodiversity from our countryside has been caused by changes in agricultural practices. It is surely evidence that we need look no further than changes in farming to explain our wildlife crisis.

But now take a look at Figure 7. This shows the same Farmland Bird Indicator over the same span of years, but this time it is plotted against the annual traffic volume on Britain's roads, measured in terms of vehicle miles. The relationship between them is just as strong as in the previous graph; in fact, in purely statistical terms it is actually a little stronger.[47]

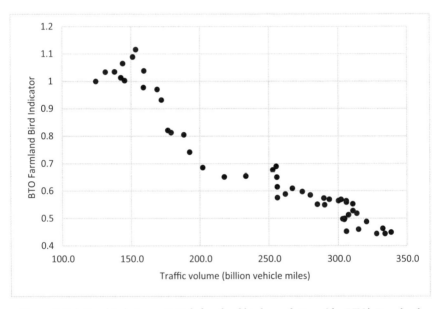

Figure 7. Relationship between British farmland bird populations (the BTO's Farmland Bird Indicator, set to an arbitrary value of 1 in 1970) and traffic volume, measured in billions of vehicle miles (data from UK Department of Transport statistics). As in Figure 6, each dot on the graph represents a different year between 1970 and 2020.

At first sight, this is a little confusing. Did we get it all wrong? Has birdsong ceased in our fields not because of changes *in* farmland, but because of changes *around* farmland? And if we believe the story told by Figure 7, what does it say about the work of all those farmland bird researchers (me included) whose results pointed the finger of blame firmly at agriculture?

In fact, both graphs are slightly deceptive. The numbers they show are not wrong, but presenting a relationship between two sets of measurements in this way easily misleads people (seasoned scientists included) into assuming that changes in one occur *because* of changes in the other. Both cereal yield and vehicle miles have increased steadily since 1970, and farmland bird numbers have declined equally steadily over that same period, but this doesn't *prove* that the former are the cause of the latter. Researchers have a little rhyme to remind themselves of this statistical pitfall: correlation does not imply causation.

But I think we should not entirely dismiss Figure 7 as a spurious correlation because it shows something interesting: just as with cereal yield, the relationship is plausibly proportional – a more than doubling of traffic volume over the last 50 years has been accompanied by a more than halving of farmland bird populations over the same period. This certainly doesn't constitute evidence of causality, but it at least *allows the possibility* that increases in road traffic have been partly, perhaps even largely, responsible for the loss of Britain's wildlife. Had the dots in Figure 7 been randomly scattered across the graph like stars in the night sky, I would never have considered writing this book. And had the relationship between the two variables not been so directly proportional – if, for instance, a doubling of traffic volume had coincided with a mere 10 per cent drop in the Farmland Bird Indicator – then I would have assumed that the impacts of road traffic have been negligible at most. Instead, there is a striking pattern that is *consistent* with a new explanation for the loss of Britain's wildlife, even if it is not *proof* of it.

Thankfully, there are two rather better lines of enquiry that we can follow if we want to assess whether or not traffication has led directly to a loss of biodiversity. First, we can look at how the abundance of different species changes with distance from the road. If we find that, all other things being taken into account, numbers of animals are reduced close to roads and recover as we move away from them, then it's hard not to conclude that roads reduce populations directly. Second, we can look at what happens to wildlife populations in those rare cases where traffic levels change but everything else stays the same.

But before we examine these two lines of evidence, we need to equip ourselves with a handy little calculator.

The traffication ready reckoner

Maps 1 to 4 (pp. 36–39) show how much of Britain falls within 100 m, 500 m, 1 km and 2 km of a road. So far, you have had to take my word for it

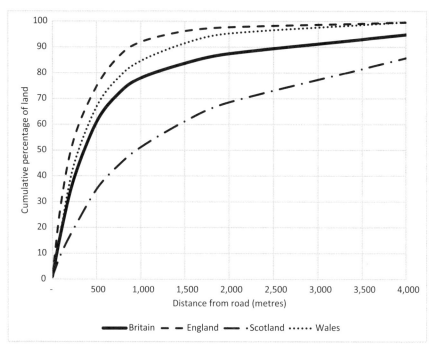

Figure 8. Cumulative percentage of total land area of Britain, and of England, Scotland and Wales separately, falling within different distances of one or more roads. The way I generated the data to produce this graph slightly under-estimates the pervasiveness of roads for the two reasons given in the caption of Map 1 (p. 36).

that the shaded areas in these four maps cover between 20 per cent (in the case of Map 1) and 90 per cent (Map 4) of Britain's total land area. But what if we wanted to know what proportion of Britain's land area falls within other distances of a road – 350 m, say, or 3.5 km? And what if we wanted to pose the question the other way around – to ask within what distance of the nearest road a certain percentage of our land falls?

We can answer these questions using a simple graph. Figure 8 shows how much of Britain's land area lies within different distances of the nearest road. Distance runs across the bottom of the graph, and the vertical axis indicates the percentage of Britain's land area falling within that distance of at least one road. The beauty of this graph is that we can use it to tell us what percentage of the country's land falls within *any* specified distance from the nearest road (though for reasons of clarity I have truncated it at 4 km). If we follow a line vertically up from a distance of 1,000 m to where it hits the solid curve, then read horizontally across to the vertical axis, we find that 78 per cent of Britain's land is within 1 km of a road – this is the shaded area shown in Map 3 (p. 38). The graph also allows us to estimate these numbers for each country separately; for road-dense England, 92 per cent of land falls within 1 km of a road, whereas for Scotland, with its extensive areas of roadless highlands, it is just over 50 per cent. The lines for England and Wales close in on 100 per cent near 4 km, meaning that almost everywhere in those two countries is less than that distance from a road, but if you were dropped at a randomly selected point in Scotland, there is still a 15 per cent chance you would need to walk more than 4 km to reach the nearest tarmac.

We can read the graph in the other direction too: for instance, we can use the graph to work out within what distance of a road 72 per cent of all Britain's land falls: the graph tells us that it is 750 m, and in England alone it is just under 500 m. And why choose 72 per cent for this example? Because that is how much of Britain's land is farmed.

This is leading towards a crucial point. It might seem inconceivable that roads, which occupy just 1 per cent of our land area, could have an effect on our wildlife that is anything like comparable to the intensification of agricultural land, which covers 72 per cent. But what the graph tells us is that if traffication reduces wildlife populations up to 750 m from the roadside, then the total area of land being affected by roads across Britain is the same as that being impacted by agricultural intensification. In England, with its dense road network, traffic needs only to reduce wildlife populations for 500 m on each side for it to affect the same total area of land that agriculture does.

The same pattern of saturation by roads holds true if we consider only areas of particular importance for wildlife. The most important wildlife sites in Britain, in terms of their international conservation significance, are called Special Protection Areas (for birds) and Special Areas for Conservation (for other animals and plants). Together they form Britain's contribution to a Europe-wide network of key wildlife sites called Natura 2000 (a name that now surely needs to be updated). Of our 770 terrestrial Natura 2000 sites, over 70 fall entirely within 500 m of a road, 260 lie entirely within 1 km of a road and nearly 500 (65 per cent) have no part of their land any further than 2 km from a road. Another designation of areas important to wildlife, called Sites of Special Scientific Interest (SSSI), adds a much larger number of sites of national or local interest to the internationally important Natura 2000 sites. Of the approximately 8,500 terrestrial SSSIs in England, around half lie entirely within 500 m of one or more roads, and well over three-quarters fall entirely within 1 km. Only a few hundred sites contain any land at all that is more than 2 km from a road. What this means is that our premier wildlife sites, the crown jewels of our natural heritage, are just as heavily trafficated as the rest of the country.

Suddenly, it does not seem quite so implausible that tarmac and traffic could be every bit as important a cause of our country's wildlife crisis as pesticides and monocultures.

The road-effect zone

In the half-century following the publication in 1925 of Dayton and Lillian Stoner's article – the first ever to highlight the risks posed to wildlife by cars – practically everything written on the subject focused on the issue of roadkill. But in 1973, a study was published that would mark the start of an entirely new direction of research. The problem was that almost nobody noticed it.

There is a famous story in the history of science about Gregor Mendel's research in the 1860s on the basics of genetic inheritance in pea plants. His groundbreaking work did not achieve the instant recognition it deserved largely because he chose to publish his results in a journal that few of the leading scientists of the day had even heard of, let alone read. Poor Mendel has been dubbed 'the patron saint of low-impact scientific publishing'. To be fair, he did send copies of his articles to some of the leading scientists of the day, including Charles Darwin (although there is no evidence that the great man ever read it). As it was, Mendel's findings lay buried in the pages of the *Verhandlungen des naturforschenden Vereines in Brünn* (Transactions of the Natural Research Association of Brno) until 1900, more than a decade after his death, when they were simultaneously 'rediscovered' by three different scholars and lit up the scientific world, instantly creating the science we now call genetics.

In a slightly lesser way, it was perhaps because a biologist called Jan Veen chose to publish his article 'The Disturbance of Meadow Bird Populations' (1973) in an obscure Dutch journal of urban planning that it went largely unnoticed by ecologists at the time, and indeed remains almost unknown to this day. However, it can stake a strong claim to being the most important publication on the environmental impacts of road traffic since the Stoners fired their starting pistol half a century earlier, because what Veen proposed was that populations of birds could be significantly reduced not just in the immediate vicinity of roads, but also for considerable distances on either side. His study of five wading birds – lapwing, ruff, black-tailed godwit, redshank and oystercatcher – nesting in Dutch meadows showed that numbers were significantly reduced for distances of over half a kilometre either side of quiet country roads, and for more than 1.5 km either side of busy roads. Veen's findings were confirmed a few years later when a more extensive study re-examined his field sites and included additional ones. Within the zones affected (which the later researchers extended to 625 m from quiet roads and 2 km from busy ones), numbers of breeding waders were reduced by up to 60 per cent. A quick check on our calculator (Figure 8) for the lower value of 625 m shows that, if this pattern were consistent in time and space, then populations of these species would be impacted by road traffic across nearly 70 per cent of Britain's land area, and 85 per cent of England's. Indeed, our populations of all five of these species have declined greatly over the last 40 years.

Here, then, was the first suggestion that road traffic could drive down the numbers of wild animals deep into the surrounding countryside. The American landscape ecologist Richard Forman, who has done much to promote the science of road ecology (not least by giving it a name), calls this band of destruction the *road-effect zone*. Some road ecologists use this term to refer to the zone of influence of roads in any regard, such as changes in behaviour or stress levels, but I will use it here more strictly to refer to the zone within which numbers of animals are reduced.

The road-effect zone is not a fixed, neatly parallel-edged band of environmental damage that affects all species in the same way. Instead each species has its own zone, determined by its sensitivity to the different pressures of traffication and the extent to which they spread outwards from the road into the surrounding countryside. But all the habitats within a species' road-effect zone, no matter how pristine in other regards, no matter how strictly protected or how well managed for conservation, are degraded or even rendered completely uninhabitable by the effects of the road.

Between 1991 and 1997 another Dutch research team, this time led by Rien Reijnen and Ruud Foppen, published a series of scientific articles that examined in greater detail than any previous study the relationship between road traffic and bird populations. These articles were published during the height of the scientific feeding frenzy of research on farmland birds, and I have always felt that, like Jan Veen's research before them, their work did not receive the attention it deserved. Reijnen and Foppen found that well over half of all the woodland bird species on their study sites occurred in lower numbers near roads. For some species the effect was fairly local, their numbers reduced only within a band of 50 m or so from the road, but the more sensitive species were significantly reduced in numbers for almost 3 km each side of busy highways. Within a fixed zone of 250 m either side of a road (which our calculator tells us would include about 40 per cent of Britain's land, and 60 per cent of England's), numbers of birds were reduced by between 20 per cent and 98 per cent, depending on the species.

Focusing on the willow warbler, Reijnen and Foppen also found that birds living near roads tended to be younger, less experienced individuals, and that they arrived on their territories later in the breeding season than did the more mature birds breeding further away. Young birds tend to be forced into less suitable habitats by older and more socially dominant adults, which occupy and defend the best sites, so this observation provided further evidence that habitats near roads are degraded, at least as far as willow warblers are concerned.

The results of these pioneering Dutch studies have been replicated many times since, in many different countries and habitats. For example, a study that looked at five species of grassland birds at different distances from roads around Boston, USA, found that low-traffic roads (carrying fewer than 8,000 vehicles per day) had no effect on their populations, but roads carrying from 8,000 to 15,000 vehicles a day reduced bird numbers up to 400 m from the verge. The road-effect zone expanded to 700 m either side of roads carrying from 15,000 to 30,000 vehicles and a massive 1,200 m either side of a multilane highway carrying over 30,000 vehicles per day.

Here, briefly, are some further illustrations of the road-effect zone from around the world:

- A study of feeding bats in the UK found that numbers are reduced over 1.5 km from busy motorways. Another study (from Australia), undertaken along much quieter roads, found that bat numbers are lowered over distances of 120 to 890 m from the road, depending on species.
- In Swiss farmland, numbers of brown hares are reduced for at least 300 m either side of roads.
- Numbers of frogs and toads in Canada are reduced up to at least 250 m either side of busy roads, though the populations of some particularly sensitive species are depleted for a kilometre or more.
- Forest caribou (reindeer) rarely venture within 5 km of a busy highway, even if their most favoured habitats fall within this zone. Moose appear to be a little more tolerant of roads but tend to venture with 500 m of them only when attracted by road salt.
- In the Mojave Desert, numbers of tortoises are reduced for up to 230 m from quiet country roads and over 300 m from interstate highways.
- Galápagos lava lizards increase in numbers by 30 per cent with every 100 m additional distance from a road.

- In the Cherokee National Forest in the USA's Appalachian mountains, numbers of soil invertebrates are reduced for up to 100 m from narrow, unpaved forest tracks.
- Numbers of some breeding birds in grasslands in the Netherlands, including declining species such as the skylark and lapwing, are reduced for up to 1.5 km from busy roads. Tawny owls may be reduced in numbers for up to 2 km.
- Numbers of Europe's largest bird of prey, the cinereous vulture, and its heaviest bird, the great bustard, are reduced to almost nothing within half a kilometre of a highway. Similar results have been reported for the gigantic Andean condor in South America.
- Fewer forest bird species are found in isolated fragments of pine forest in Catalonia within 2 km of a major highway then in those further away. Similar results have emerged from studies of birds in Cyprus.
- A small, unmarked road running through a nature reserve in Western Amazonia significantly alters levels of species richness, diversity, abundance and community structure in birds, butterflies and amphibians for at least 350 m on either side.
- In Africa, numbers of western chimpanzees are reduced up to 5 km from minor roads and an astonishing 17 km from major roads, the widest road-effect zone yet recorded. Less than 5 per cent of the western chimpanzee's entire range in West Africa remains unaffected by roads.

There are plenty more examples, although almost all studies of the road-effect zone have been of vertebrates. We know very little about how roads affect the abundance of insects or plants (though one study from Australia suggested that certain groups of insects can show a wide road-effect zone). The road-effect zone for many roadside plants might extend only a few metres from the verge, although if they rely on other species with wider road-effect zones for services such as pollination and seed dispersal, these disadvantages may also transfer to the plants themselves. Such an effect has been recorded in South Africa's hugely biodiverse but fragile fynbos ecoregion, where scientists have found that plants pollinated by birds have lower rates of pollination for tens of metres either side of roads. In California's dry grasslands, certain native plant species only occur 1 km or more from the nearest road; any closer and they are outcompeted by invasive species spreading outwards from the roadside.

For vertebrates at least, the results are consistent and compelling – traffication drives down the populations of many species for hundreds or even thousands of metres either side of the road. An overview of many published studies concluded that numbers of larger mammals, reptiles, amphibians and birds are generally reduced by the presence of roads, whereas small mammal numbers are unaffected or even higher closer to roads (explaining the high roadkill rates of rodent-hunters such as barn owls). Another review has concluded that numbers of birds are reduced up to an average of 1 km from roads and other noisy infrastructure, and those of mammals, particularly larger species, up to an average of 5 km. Our calculator tells us that with a road-effect zone of 1 km, bird populations will have been reduced by road traffic across 78 per cent of Britain's land, an area even greater than that impacted by agricultural intensification, and over 90 per cent of England might have been affected. For larger mammals, with a road-effect zone of several kilometres, populations must have been reduced by roads across pretty much the entire country.

The most ambitious plans for Britain's wildlife go beyond simply restoring populations of recently depleted species: they aim to reintroduce some of the larger mammals and birds that were driven to extinction centuries ago – lynx, beaver, great bustard, wolf and bison – and to recreate lost landscapes and wildlife communities. How wonderful that would be, but these rewilding projects will need to face the reality of traffication, because large-bodied, slow-breeding species are usually those with the widest road-effect zones. The ambitious conservationists behind these laudable restoration projects all clearly foresee the potential for roadkill to thwart their ambitions, but fewer may recognise that this might not be traffication's only obstacle to their success.[48]

The most powerful and convincing of these road-effect zone studies are those that simultaneously take account of other possible causes. My former PhD student Sophia Cooke tackled this issue using the huge dataset of the BTO's Breeding Bird Survey. In line with previous studies, Sophia's results showed that many species show a fall in numbers with increasing exposure to roads, and for road-sensitive species the average effect distance is 700 m. This is the best estimate of the road-effect zone we have for birds in Britain; our calculator shows that these declines will have occurred over 70 per cent of Britain's land, and that around 85 per cent of England will have lost significant numbers of some species to road traffic.

What makes Sophia's analyses so powerful is that she built the effects of agricultural intensification into her calculations, together with other factors that might influence bird numbers, such as habitat type, human population density and climate. Furthermore, she also accounted for the fact that the people who collected the data in the field might have been less able to detect birds near roads, owing to the noise of traffic. The influence of roads remained strong when the effects of all these other variables were factored in. Intriguingly, she also found that long-distance migrants, a group of birds in severe and protracted decline for reasons that are still unclear, showed a significantly greater avoidance of roads than species that do not migrate. We shall return to this very important result in Chapter 10.

Taken together, this wealth of studies provides compelling evidence that roads reduce the populations of many animals across great swathes of our countryside. What factor other than traffication could possibly explain why numbers of animals are so consistently low near roads, then gradually recover as the distance from the roadside increases, until they plateau out hundreds or even thousands of metres away? Wheatfields are not managed more intensively closer to roads than further away from them, so that cannot explain the pattern. Other widespread threats, such as global warming, are equally unlikely to affect areas close to roads more than those further away. Nor is it the case that the lower numbers of birds counted near roads is simply because the people counting them are deafened by the noise of traffic, because several studies, Sophia's included, account for this in their analyses. These patterns can only be due to the direct impacts of road traffic.

The anthropause

The second strand of evidence we can use to assess the impacts of traffication on the abundance of animals comes from studies that examine those rare situations in which road traffic changes but everything else, such as agriculture and climate, stays exactly the same. Evidence like this has proved compelling in proving a causal link between threats such as agricultural intensification or invasive species and declines in biodiversity – reverse them by changing how land is farmed, or by eradicating pest species, and the native wildlife usually recovers.

In the case of traffication, evidence of this type is rare, for obvious reasons. It would be scientifically wonderful, although of course massively inconvenient to millions of people, to close some motorways for a few years and see how the wildlife in the surrounding landscape responds, but that is unlikely ever to happen. Such evidence in the field of road ecology is therefore largely opportunistic. There are not many examples, and none that I can find from Britain, but the few studies that exist are all entirely consistent with the existence of a wide road-effect zone.

A study carried out in Spain, for instance, looked at how birds of prey changed their distribution in response to changes in traffic flow along a major road. At weekends, when the number of vehicles more than doubled compared with weekday traffic flow, vultures and eagles moved significantly further away from the road to feed and roost – a considerable area around the road was effectively lost to them. The researchers chose their study site well: the road was not used by pedestrians, and all the surrounding land was privately owned, so the results could not have been caused by a weekend influx of people moving about outside cars. Similar results come from Canada, where the closure of a popular road in Banff National Park during the hours of darkness allowed grizzly bears and other mammals to feed closer to the highway at night than they had done previously.

But the greatest unplanned experiment of all time was that which descended upon humanity in early 2020. The COVID-19 pandemic and the lockdowns put in place to control its spread led to the first major reduction in road traffic since the Second World War. The world went into what has been called an *anthropause*, a temporary relaxation of the pressures we place on the environment. Highways and motorways briefly fell silent, as traffic dropped to levels not seen since the 1950s. This gave scientists a unique opportunity to see what happens when traffic is reduced over huge areas. Birds and frogs immediately changed their songs. Roadkill rates plummeted. And using millions of observations of birds submitted by volunteers, researchers in Canada and the USA were able to show that many bird species that normally avoid roads were found in significantly higher numbers near them during lockdown than in previous, non-lockdown years. A single spring of calm is not long enough to make much difference to the overall populations of any species, but the fact that these birds settled much closer to highways during the brief COVID-19 anthropause suggests that these areas remain suitable for them in every other way, and that they have been abandoned solely because of road traffic.

Phantom roads

The scientific evidence is overwhelming: road traffic drives down populations of wild animals deep into the surrounding countryside. The effects of traffic permeate our road-dense environment so extensively that few areas are unaffected. Roads affect our wildlife populations just as pervasively as do agricultural intensification and climate change. Traffication therefore deserves recognition as one of the leading causes of our country's precipitous loss of biodiversity over the last half-century. When we understand the problem better, it might even prove to have been the *most* important.

But we have seen already that trafficcation is not one threat: it is a composite of many. Which of the different elements of trafficcation is most responsible for creating the road-effect zone? This has been a matter of some debate among road ecologists. Some have argued that roadkill is the main problem, slaughtering huge numbers of animals on the roads and creating a vacuum of unoccupied habitat that sucks in others from further afield to meet the same fate, in much the same way that drawing too much water from a well can lower the water-table for hundreds of metres around. This might well be true in some cases, but we have seen already that the commonest victims of roadkill are usually the species that occur in highest numbers around highways. For many of these species, the road-effect zone is *inverted* – their numbers are highest around roads and lower further away (we will return to these road-tolerant species in Chapter 10). This is not the pattern we would expect to see if roadkill were the main problem.

Instead, a consensus is now emerging that for many species, particularly mammals and birds, the main cause of the road-effect zone is probably traffic noise. We have seen how noise from even minor roads can spread deep into the countryside to affect sensitive animals over huge areas. Several studies of the road-effect zone have included measurements of noise in their analyses and found that this explained the observed patterns better than anything else. Laboratory experiments show the same thing: when presented with a choice, animals move away from traffic noise and into quieter areas.

Environmental scientists do not have the money or influence to close major highways and compare what happens in newly de-trafficated landscapes to those in which traffic levels are not changed. But they can do the opposite – create artificial highways in road-free environments. A remarkable experiment carried out in Idaho, USA, in 2012 did exactly this. Here, researchers created a 'phantom road' in an area of undisturbed forest by setting up a line of loudspeakers half a kilometre long through which they played the noise of moving traffic. The level of road noise they broadcast through the speakers corresponded to a fairly low volume of cars travelling at modest speeds of around 45 mph, because the aim of the research was to mimic typical traffic flow through a national park. Counts of birds were made along the phantom road before, during and after the sound was broadcast, and at undisturbed comparison sites nearby.

The results were striking. Despite the modest simulated traffic volume, over half the species originally present showed significant falls in numbers along the phantom road while noise was being played. The overall number of birds declined by nearly a third, and some species disappeared completely. And as with Reijnen and Foppen's willow warblers in the Netherlands, the researchers found that the birds that remained near the phantom road tended to be young, inexperienced birds, forced into these disturbed habitats as adult birds moved out to occupy undisturbed forest elsewhere. A second phantom road experiment, this time on the island of Hokkaido in Japan, yielded very similar results: the numbers of woodland birds dropped when traffic noise was broadcast through loudspeakers (although in this study there was no effect on grassland birds).

These phantom road experiments elegantly demonstrate that traffic noise on its own is sufficient to create a road-effect zone for birds, and they show just how sensitive some species can be to even fairly low levels of soundscape pollution. Traffic noise causes a wide band of otherwise perfectly good habitat to be effectively lost, and populations fall.

A further line of evidence to justify the noise hypothesis is that species of birds with lower-pitched songs, which overlap more with the noise of traffic, tend to have wider road-effect zones than species with higher-pitched songs (the 'piccolos' we encountered in Chapter 6). Research on woodland birds in Germany found that species with higher-pitched songs, their frequencies above those of traffic noise, have narrower road-effect zones than species with lower songs, showing that they are less affected. Similar results have been found in studies of birds in the USA and elsewhere; the species least able to withstand the impacts of road traffic are those whose songs overlap most with the noise frequencies of road traffic, suggesting again that the most destructive force of traffication, for birds at least, is noise.

The picture is becoming clearer. Traffic noise creates a soundscape of fear, causing animals to abandon habitats that are otherwise perfectly suitable. Ecological factors dictate that these displaced animals cannot simply move into areas further from roads that are already occupied by others of the same species – there are only enough resources in these quieter areas to support the animals that are already there. The only possible outcome of these evictions from huge areas of otherwise suitable habitat is that populations fall across the landscape.

This does not mean that all the other impacts of traffication – air and light pollution, fragmentation, roadkill, salination and so forth – have not also played a part in our country's biodiversity loss. It is very likely that these additional hazards exacerbate the impacts of noise, further widening the road-effect zone, and that they increase the vulnerability of wild species to other threats. For insects and plants, factors other than road noise are probably far more important in determining their road-effect zones. As climate change starts to bite, fragmentation by roads might become a far bigger problem than traffic noise.

There is a troubling corollary to this, which is that much of what we know about our wildlife is based on field research that has been undertaken in the shadow of traffication. Taking just British birds as a subject, literally thousands of research papers have been published in the last few decades on subjects as varied as nesting ecology, diet, feeding behaviour, survival and mortality rates, migration, habitat selection, genetic patterns, health and disease, breeding behaviour, population dynamics, conservation needs, vocal communication, plumage patterns and so forth. Much of the data gathered for these studies has inevitably come from areas suffering different degrees of traffication – if we assume a road-effect zone for birds of a kilometre, we only need to look again at Map 3 (p. 38) to see the truth of this. But we are now starting to realise that road traffic can affect pretty much every aspect of avian life, from blood stress hormone levels to song behaviour to nesting success to survival to population dynamics. So most of what we know about our birds, or rather what we *think* we know about them, is heavily influenced by something that the great majority of ornithologists simply ignore, and each study will be skewed to a different degree. We may know almost nothing about how birds live and die in an untrafficated world.

Apportioning the blame

Let's put science aside for a moment and indulge in a little fantasy. Imagine we have four identical copies of Britain as it was in 1960 and the power to control their destinies. The first of these four worlds we will treat as our control. Here, we freeze both agriculture and road traffic exactly as they were in 1960, and then let the country run its course to today. We rather implausibly allow everything else – climate change, politics, industry, population growth, advances in medicine, air travel – to proceed exactly as they have done. Computers appear on every desk, smartphones in every pocket, but our roads and fields remain suspended in time. In the second copy of Britain we do the same except that this time only agriculture is frozen at its pre-Beatles level, and we allow road traffic to run its unnatural course along with everything else. In the third copy, we reverse the treatments; here, agriculture follows its path towards intensification but our road traffic remains fixed as it was in 1960, Morris Minors and all. In the fourth and final copy, we allow both agriculture and road traffic to progress as they have, alongside everything else; it is, of course, the world as we know it.

Now try to imagine a journey through each of these four worlds today. In the first, we would drive past ancient hedgerows that enclose a varied patchwork of fields growing a mix of different crops in rotation. Many of the cereals are spring-sown crops, bloodied with poppies, that provide opportunities for ground-nesting birds in summer and stubble fields full of grain in winter. Grass crops (leys) rotate around the farm like cereals, so meadows and arable land are all jumbled up together. Flower-rich hayfields are cut just once a year at the end of the summer, after grassland birds have finished nesting. Our journey through this world is easy, for there are few other vehicles on the road. The road is not straight and our car not built for speed so we drive slowly, but even at this sedate pace our windscreen is soon pockmarked with the bodies of insects, little stars of coloured ichor smeared into rainbows by the wipers. As we get out to clean the glass, we pause and look around. The sky is crazed with skylark song, and like Edward Thomas at Adlestrop station we fancy we can hear birds singing from counties around. Along the field edges and blending into the crop are carpets of flowers whose names reveal their old arable affinities; corncockle, corn marigold, corn buttercup.

In the second world we are keen to pause and look over the hedge, wanting to see what wildlife lies on the other side, but stopping here is dangerous, for a modern road network runs through an old landscape. Each time we pull over onto the narrow verge our car is rocked by the suction of juggernauts as they barrel past us, horns blaring. We cannot safely open the door to get out, so we lower the windows and listen but the noise from passing vehicles is too much. After a while our stress levels rise and we give up. We look anxiously in our mirror for a break in the traffic and rejoin the flow, accelerating hard as headlights flash in anger behind us. We leave knowing nothing about this world's wildlife.

In the third world stopping the car is easy, for there are few other vehicles on the road, but finding a place where we actually want to pause is more of a problem. We drive around quiet and overgrown lanes, cast iron signposts pointing askew to villages that remain well off the beaten track, but all we see to either side are wheat monocultures. There must be wildlife in this quiet and peaceful land, so we park up in a broad gateway and get out to stretch our legs. As we do, a huge machine appears in the field next to us and cones of spray jet from ten-metre booms. We are not sure what this mist contains but it has caught the wind and is drifting in our direction, so we rush back to the car, close the windows and drive on. Again, we have learned nothing of the biodiversity that this alternate reality holds.

A journey through the fourth world poses no problem to the imagination, because you can do it any time you like.

There is a serious point to all this daydreaming, which is that this fanciful multiverse experiment is the only way we could calculate, precisely and beyond any doubt, the relative contributions of farming and road traffic to the countryside's loss of wildlife. The interplay of all the different environmental changes that our country has lived through over the last half-century is too complex for us ever to know what proportion of our wildlife loss we can blame on traffication, what proportion on agriculture, what proportion on climate change and so forth. We are even further from being able to break each of these broad classes of threat down into its component parts – separating out the impacts of roadkill from those of traffic noise, for example, or those of pesticides from the loss of hedgerows. But if we could, we would then be able to rank the various threats and create a league table of guilt, allowing us to estimate, for each problem separately, how much of our lost biodiversity we would get back if we fixed it.

Unfortunately, an exercise like this is so fraught with difficulty, and would require so much new (and old) data, that it has never been attempted. We will never know how many of the million-and-a-half pairs of skylarks that we have lost from Britain's countryside disappeared due to traffication, how many because of agricultural intensification, how many through a combination of the two, and how many to other causes entirely. But road traffic must surely have played a huge part – we know that even light traffic along minor roads can depress skylark populations for up to 3 km.

Scientists will need to find some other way to try to apportion the blame for our wildlife losses between the competing threats of traffication, agricultural intensification, woodland management, African drought, climate change and others, and to identify all the complex synergies between them. Solving this problem represents, to my mind, the greatest challenge for environmental scientists in the coming decades, although I suspect it may never be possible.

In the meantime, the scientific evidence that identifies traffication as a serious threat to wildlife is just as compelling, and thanks to the recent growth of road ecology just as extensive, as that which points the finger of blame at agricultural intensification or climate change. The evidence overwhelmingly tells us that roads have contributed hugely to the collapse of our country's biodiversity, even if it does not allow us to estimate exactly how much wildlife our love of the car has cost us.

There may be those who, even in the face of all this evidence, refuse to accept that driving heavy, noisy chunks of speeding metal 15 trillion miles each year over our little planet's fragile green carpet of life causes huge environmental damage. My exasperated and rather unscientific response to them is the same as that I offer to those few diehards who still refuse to accept that pumping billons of tons of greenhouse gases into the atmosphere causes climate change: how could it possibly *not*?

It might sound unlikely, but the scientific evidence is overwhelming: we are blasting our wildlife away with traffic noise, chopping it into pieces with roads, flattening it with cars and poisoning it with exhaust fumes. Traffication has sucked the life out of our countryside. We are, quite literally, driving our wildlife to extinction.

And if that comes as news to you, then you are certainly not alone. Because one of the most worrying things of all about the wildlife-strangling giant of traffication is that almost nobody seems to have noticed it.

CHAPTER 9

The Sixth Horseman

For reasons that defy any logical explanation, efforts to avert the environmental collapse of our planet, and thereby save ourselves from utter extinction, are very poorly funded. The total amount of money spent each year on wildlife conservation, from all the world's governments, private donors, charities and other sources combined, is less than the projected cost of Britain's HS2 rail scheme, which will destroy acres of ancient woodland to shave 20 minutes off a journey between London and Birmingham. We could pretty much halt global extinction and protect all the world's most important wildlife sites for a tenth of the money we spend each year on soft drinks, yet there is no slowdown in the rate at which our planet's life support systems are shutting down.[49] The lightly armed dragoons of conservation hold a thin green line around some of the world's most vulnerable species and ecosystems, but can do little to hold back a rising tide of extinction. Even the most optimistic assessments accept that things will get much worse before they start to get better, if indeed they ever do. All we can do for now is marshal the scant resources at our disposal in the most effective way possible and save as much as we can from the wreckage.

An essential starting point in any conservation effort, therefore, is to identify correctly the problems that most urgently need to be addressed and to develop effective solutions to them; this has been my work as a conservation scientist for more than three decades. Misdiagnose the problem and we will probably prescribe the wrong remedy, wasting already woefully inadequate resources and potentially making things even worse. The Hippocratic maxim 'first, do no harm' applies just as much to those trying to nurse the environment back to health as it does to medical practitioners.

If you were to ask some informed conservationists what has reduced the planet to its dire environmental state, I suspect they would all give you broadly similar answers. They would list the threats posed by habitat destruction, agricultural intensification, invasive species and the unsustainable consumption of natural resources – the four original horsemen of the environmental apocalypse. They would then tell you that more recently a fifth horseman has galloped into their midst in the form of climate change.

These dark riders form a tight and fast-moving cavalry unit and it can often be hard to see where one ends and another begins. The spurs of habitat destruction goad the flanks of climate change, since the burning of forests is a major cause of global warming. Agricultural intensification nudges habitat destruction forwards. However, the majority of recent global overviews of biodiversity loss, and most strategies to address the problem (including the international Convention on Biological Diversity), recognise these five broad categories of threat as those we most urgently need to address if we are to prevent global ecological meltdown.

But if it were possible to freeze this ecocidal cavalry charge, as in a photo finish, and examine it closely from different angles, I think that we would be able to discern in it not five apocalyptic riders, but six. Having read this far you will have no difficulty in guessing the identity of my putative sixth horseman: it is traffication.

Conservation's blind-spot

The argument I have tried to present in this book is not much different from that proposed by Dayton and Lillian Stoner a century ago. It is that in regions of the world with dense networks of roads – which is now *many* regions of the world – traffication has directly brought about a loss of biodiversity that is comparable in scale to, and quite possibly exceeds, that caused by such environmental *causes célèbres* as agricultural intensification, habitat loss, hunting and (so far at least) climate change. It is not just that cars and wildlife just don't rub along well together – that would hardly be news to anyone. Rather, the scientific evidence suggests that driving cars is one of the most damaging things we have ever done to our wildlife, and in many places perhaps the *most* damaging.

Roads are not simply narrow lines of collision danger for the occasional badger or barn owl. They create broad, overlapping bands of environmental damage that drive down wildlife populations and fundamentally alter the structure of natural communities over most of our countryside. The impacts of traffication seep deep into the surrounding landscape, like a road map drawn in wet ink on blotting paper. The populations of individual species and the structure of entire wildlife communities are altered for hundreds or thousands of metres either side of the tarmac, leaving few places unchanged. We cannot separate out the unique impacts of roads from those of other conservation problems, such as agricultural intensification, habitat loss or climate change. But the scientific evidence levelled against traffication is just as strong: the collision, fragmentation, invasion, disturbance and pollution caused by roads are severe threats to our wildlife; they are additional to other environmental pressures; and they affect most of our land. This surely makes traffication one of the biggest environmental problems of our time. Yet it remains strangely neglected.

In 1998, the American landscape ecologist Richard Forman published an influential essay entitled 'Road Ecology: A Solution for the Giant Embracing Us'. He began with a riddle: what, he asked his readers, is 'huge, conspicuous, and avoided by ecologists'? The answer of course, and the embracing giant of his article's title, is the modern road network. The point Forman was making was that despite the scale of the environmental problems caused by road traffic, few of his fellow scientists appeared to be interested in the subject. Perhaps he had looked across the Atlantic and wondered why it was that so many researchers in Europe were working to unravel the complex relationships between wildlife and agricultural change, yet so few were working on what was, in his view, an environmental problem of similar magnitude.

Ecologists clearly heard Forman's rallying cry, as the last 20 years have seen a surge of research interest on roads and wildlife. The overwhelming majority of the hundreds of scientific articles and reports I have amassed while researching this book have been published since Forman's article appeared, and annual output is rising steadily. This new branch of study has attained sufficient gravitas and momentum to gain recognition as a discrete discipline with its own name, taken from the title of Forman's essay: road ecology. The discipline of road ecology is now a quarter of a century old and there are dedicated research groups at a number of universities around the world; conferences are held on the subject and a host of eager doctoral students are treading the tarmac to collect data. It is a field of research that remains dominated by North American scientists; in Europe, the charge has been led largely by Dutch, Iberian and Scandinavian researchers. This boom of interest in road ecology has, however, almost entirely bypassed the UK, for reasons that utterly elude me.

Scientists have risen to the challenge, but others have not. With the change of a single word I think that the riddle Forman posed in 1998 is still valid, indeed urgent, today: what is huge, conspicuous and avoided by *conservation*? For despite Dayton and Lillian Stoner's warning of a century ago, and the mass of scientific evidence published since, roads remain strangely absent from our conservation road map.

A wealth of recent scientific research has unequivocally identified road traffic as a significant cause of biodiversity loss, but you will search in vain through the strategy documents of our largest conservation organisations to find a single mention of it. There may be a big furore when a proposed new road threatens to destroy an important wildlife site, but the problem is generally viewed as a purely local one. After all, isn't the environmental impact of a road simply the replacement of a relatively small area of fields or woodland by the same small area of tarmac? Surely roads are no more of a threat to our wildlife than a new shopping centre or a housing estate?

A report published in 2003 on the possible causes of the UK's collapsing woodland bird populations, a document that helped to set the research agenda on this issue, exonerated road traffic at the outset on exactly this flawed premise: 'It seems very unlikely that traffic could be a major contributor to the declines in woodland birds because all the relevant species show declines in wooded areas away from major roads.' While I was researching this book, I discussed it with a number of friends and colleagues who are world-renowned ecologists and conservationists, and their average response can be summarised as: 'How can you write a whole book about a few flattened hedgehogs?' Conservationists see the car as a way of getting to work, but clearly not yet as a reason for coming to work.

But the scientific evidence screams a different story. It yells at us that the impacts of traffic burn deep into the surrounding countryside, that even a quiet country road can drive down wildlife populations for a mile or more around, can cleave populations apart and disrupt evolutionary processes, can fundamentally affect our wild species' ability to respond to climate change and break down vital animal communication systems. In our heavily trafficated country, pretty much everywhere has been affected.

Of the bird-centred conservation organisations perhaps only the Barn Owl Trust, the gorgeous subject of whose devotions is particularly vulnerable to collision with vehicles, flags up road traffic as one of its main concerns. Froglife and other amphibian conservation organisations have long recognised that roads pose a huge threat to their charges, and others are becoming increasingly concerned by research that implicates road traffic in declines in bat populations. But among the larger mainstream environmental organisations, in Britain and elsewhere, and among the public generally, road traffic is simply not seen as one of the important environmental issues of the day. If it gets any attention at all, it is usually for its contribution to climate change; this is hugely important, of course, but it is only one aspect of traffication's damage.

Here are just a few examples of the conservation movement's collective road-blindness:

- Wildlife and Countryside Link, a coalition of over 50 UK conservation organisations with a combined membership of over 8 million people, has specialist groups on issues such as agriculture, wildlife crime and invasive species, among others. But nothing on roads.
- The International Union for the Conservation of Nature, which unites nearly 1,500 conservation and other wildlife-linked organisations around the world, supports a huge number of specialist groups, including expert panels on oil palm plantations, pesticides, climate change, invasive species, trade in wild birds and many other issues of global environmental concern. But, again, nothing on roads.
- Reviews of nature conservation in Britain published over the last decade have discussed problems such as the persecution of birds of prey, the afforestation of the Flow Country, the reclamation of coastal wetlands, the collapse in sand-eel populations and their consequences for seabirds, climate change, the problems of introduced non-native species, the expansion of housing and industry, changes in woodland management and, of course, agricultural intensification. But I have yet to see one that mentions the rise in road traffic.
- Recent books, reports and scientific articles on the collapse of bird populations in Britain, across Europe and across North America identify climate change, habitat loss, agricultural intensification (particularly pesticide usage), invasive species, persecution and collision with buildings as the main causes. But not one makes mention of roads.

It is almost as though the hundreds of scientific articles that I have tried to distil into the previous chapters were never written, or at least never read.

But there is perhaps a good reason for our collective road-blindness, one that I think has not been spotted before.

The importance of names

To an ecologist, if not to a farmer, *agricultural intensification* is a catch-all term that is used to summarise a number of related threats to biodiversity that manifest themselves in different ways. Sitting beneath this banner headline are the risks posed to insects and wild plants by agrochemicals, the loss of hedgerows and ponds, the creation of monocultures, the destruction of birds' nests through earlier crop harvesting and so forth. One of the great insights of the 1990s boom in farmland biodiversity research was that agricultural intensification is not one single ecological threat but many quite different ones, all bound together by a common mechanism.

The same is true of the term *climate change*, which encapsulates a number of quite different threats to wildlife that are unified by the common causal mechanism of global warming. It includes the problems faced by species that need to shift their ranges to track moving climates, disruptions to the timing of biological events, the mass mortality of animals through droughts, heatwave and fires, the increased spread of invasive species and many other incipient (and indeed ongoing) catastrophes.

The enormous cumulative threat posed to wildlife by the burgeoning of road traffic is also a collection of different threats. It includes roadkill, the pollution of soundscapes, water, air and darkness, the fragmentation of landscapes into traffic islands and the associated loss of genetic diversity, the accelerated spread of invasive species and all the other many other problems discussed in the preceding chapters. Exactly like agricultural intensification and climate change, it is a grouping of many quite different environmentally damaging processes that are linked together by a common causal mechanism.

But there is one hugely important difference: *it doesn't have a name.*

There is no widely accepted term that encapsulates and summarises the growth in the number, speed and polluting power of vehicles or the distances they travel, the expansion (lengthways and widthways) of our highways, and the increasing pervasiveness of road traffic in our countryside and in our lives. In its absence, I have had to contrive the word *traffication*, because I could not have written this book without it. It seems quite extraordinary to me that one of the most profound societal and environmental changes of the last century lacks a common term of reference.

This is not just a semantic point – names matter. The many disparate threats that are subsumed under collective terms like *agricultural intensification* or *climate change* have different ecological impacts that often require very different conservation solutions. But by recognising their interconnectedness and grouping them together, the whole becomes greater, and more recognisable, than the sum of its parts. A wide range of hugely complex processes can be captured in a way that people can recognise and relate to, even if they don't fully understand the intricacies of what lies under the hood. By linking them together, the entwined strands of change become a *thing*, generating a gravity that pulls people and resources together to address the problem. These terms are the seeds around which crystallise research teams, stakeholder groups, policy units, publicity campaigns, funding bids and all the other paraphernalia needed to change the world.

At the moment, we lack a word to describe a phenomenon that links road-killed tapirs in Brazil (p. 56), shouting white-crowned sparrows in San Francisco (p. 104), short-winged cliff swallows in Nebraska (p. 59), genetically divided populations of wood mice in Spain (p. 76), the unstoppable westward march of cane toads across Australia (p. 80), poison-bearing honeybees in Italy (p. 121), the mass die-off of salmon in the Pacific (p. 123) and the salination of North America's lakes (p. 125). But they are all products of the same underlying change in the global environment.

This peculiar omission from our vocabulary is more than just another illustration of our collective road-blindness; it may go a long way towards explaining it. Because if we have no word for a problem, how can we even begin to talk about it, let alone respond to it? In 1935, the American wilderness pioneer Robert Marshall bewailed the fact that 'The arguments in favour of roads are direct and concrete, while those against them are subtle and difficult to express.' Thanks to the growth of road ecology, this is not as true today as it was in 1935 – the arguments against roads are now very much more clearly defined and supported by a wealth of scientific data – but just think how much easier it would be to make these arguments if we had a word to describe the problem.

I have no satisfactory explanation for why no word exists in our lexicon to describe something that has so fundamentally shaped our lives and our environment, for better and for worse, for more than a century. Perhaps it is because traffication started long before debates around climate change and agricultural industrialisation, and has proceeded more slowly. Of course, the world's climates have been changing for as long as they have existed, and agricultural yields have done nothing but increase over the 12 millennia or so since wild grasses were first domesticated. But the terms as they are used now apply explicitly to the rapid 'hockey stick' changes that started only 30 or 40 years ago. Cars and roads, on the other hand, narrowly pre-date the modern conservation movement.

Only a relative handful of people alive in the world today have experienced the world as it was before the advent of mass motoring, and then only in regions to which traffication came late. Vehicle numbers, the distances and speeds they drive, and the length and capacity of our road network have all increased hugely over the last century but their growth has been slow and steady, just 1 or 2 per cent each year. Perhaps we simply did not notice the slow increase in traffication and the gradually rising environmental costs that it has brought. If so we have been negligent, because as early as 1911 an anonymous Londoner had spotted the dangers of creeping traffication:

> If one quarter of the changes of street traffic that have happened in the last ten years had come upon London suddenly, they would not have been tolerated; but the changes have been so gradual, the nuisances have been so wonderfully mingled with benefits, and the whole system of traffic so greatly accelerated, that the increase in noise passed almost unnoticed.

My friend Jeremy Wilson, an old comrade-in-arms from the farmland bird research boom of the 1990s and now head of the RSPB's research department, has suggested to me that our collective blind spot for roads might be an example of the boiling frog metaphor. A frog dropped into hot water, the story goes (and experiments cruelly confirm), instantly senses danger and tries to jump out, whereas a frog placed in cold water that is then slowly heated does not detect the rising threat and is boiled alive. The boiled frog, perhaps, explains all the billions of squashed frogs.

A creeping threat like traffication has no Chernobyl moments, no head-line-grabbing heatwaves, floods or wildfires to thrust it into our collective consciousness. There has been no automotive equivalent of *Silent Spring*, nothing to set a car-shaped boulder rolling down the loose scree of public concern. We see plenty of dead animals lying on the roadsides, of course, but they have been lying there all our lives. Rather than alerting us to the danger, their constant presence and familiarity might have had exactly the opposite effect, desensitising us to the threat of traffication and allowing it to hide in plain sight.

Traffication is not a new threat to our wildlife; it is an old one that somehow got forgotten before it could even acquire a name. Dayton and Lillian Stoner sounded the alarm bell a century ago, but few heeded it.

A first message of hope

Perhaps another reason for our road-blindness is that this is an environmental problem we are all responsible for, and it is never easy to point the finger of blame at ourselves. The same problem does not apply quite so directly in the case of agricultural intensification or climate change: most of us are drivers, but most of us are not farmers or oil producers (although we are all willing, even captive, consumers of their products).

Perhaps also we see our reliance on the car as being too deeply ingrained in society ever to yield to influence; no politician or wildlife charity wants to arouse the anger of the transport lobby, which, if defined as drivers of cars or consumers of road-transported goods, is pretty much all of us. Our willingness to accept the car and all its faults appears to be unwavering. The author of *Man and the Motor Car*, a largely pro-car publication, lamented in 1936 that 'we have shown little fighting spirit in the face of the hazard that the automobile has created'. Campaigning to restrict car use may not endear large environmental organisations to all of their members, and might dissuade others from joining. Furthermore, one of the benefits of vehicle ownership – indeed, the very reason that motoring took off in the first place – is that it grants urban dwellers access to the countryside, and it is well known that people need to experience wildlife if they are to feel moved to protect it.

If this is why conservationists do not engage more in this arena, I think they are misguided. *Silent Spring*, Rachel Carson's famous and hugely influential exposé of the perils of pesticides, did not conclude with a call that they be banned, simply that their use be moderated and their impacts more widely understood. And nobody has ever seriously suggested that the solution to the environmental problems of agricultural intensification is to outlaw farming. Modern crop monocultures rank among the world's most hostile wildlife habitats, repeatedly doused in biocides and scalped by mechanical shears. But research has shown the way; we can subtly tweak the management of these fields to make them much less hostile to birds such as the skylark and at the same time reward, or at least not punish, the farmer. There is no reason we cannot now do the same with our roads. Indeed, if we reduce the environmental impacts of roads, all the evidence suggests the natural health of our farmland will improve as a consequence.

Conservationists might argue that their hands are full. There are more than enough huge problems for them to deal with already, and the last thing they need to hear is that there's another one sitting right outside in the car park. Having worked in this underfunded, overstretched sector for all of my career, I can sympathise; the barrage of environmental threats that require immediate solutions is overwhelming.

But let me offer a message of hope – the first so far in this hitherto rather doom-laden book, but not, I promise, the last. If we can embrace the reality that the loss of our wildlife has resulted at least partly, and perhaps even largely, from traffication, then putting nature back into our depleted countryside suddenly becomes a lot easier than we might currently think.

Traffication casts an invisible blanket of environmental harm over our countryside that adds to all the other, more widely recognised, causes of biodiversity loss. Species that are already known to suffer from the impacts of agricultural intensification, such as the skylark, grey partridge and lapwing, are also known to be highly sensitive to the impacts of roads; their road-effect zones overlay an already bird-depleted checkerboard of chemical-dependent monocultures. Each of these two processes takes its own toll on wildlife, and even though we cannot apportion blame between them it is important that we understand that their effects are separate and probably mutually reinforcing. Reducing one of these problems will certainly allow our wildlife to cope better with the other.

The reason that extensive measures to improve the quality of agricultural lands for wildlife are not achieving their hoped-for objectives might be because the problems of traffication are not being addressed at the same time. Huge amounts of money and effort have gone into developing solutions and paying farmers to implement what are called agri-environment schemes, but the BTO's Farmland Bird Indicator, although no longer plummeting, obstinately refuses to move upwards. Perhaps traffication is keeping it down; to restore to our countryside the levels of biodiversity it supported in the 1960s we may need to address both problems at the same time. Efforts to improve conditions for woodland birds or for migratory birds might be similarly thwarted by the ever increasing pressures of road traffic. There is already good evidence that migratory species are particularly susceptible to traffic noise: perhaps their problems lie closer to home than we realise.

Take, for instance, the cuckoo. This classic harbinger of spring is one of Britain's most rapidly declining birds, and it has more or less disappeared from great swathes of our countryside. But these declines have not been uniform across the country; England has lost nearly 70 per cent of its cuckoos, particularly in lowland areas, but Scotland's cuckoo population is doing much better; indeed, it is even increasing in places. The reasons for this remain something of a mystery, particularly since many of the cuckoo's most frequently exploited hosts, such as the dunnock, pied wagtail, robin and reed warbler, remain common and some have increased in number (perhaps as a result of traffication, as we shall see in the next chapter).[50] Several theories have been advanced to explain the north–south pattern of gain and loss. Clever researchers have fitted tracking devices to cuckoos, showing that Scottish and English cuckoos follow different migration routes to Africa, one of them more prone to drought than the other, so maybe this is what underlies their different population trajectories. Cuckoos have also been hit by declines in moth populations brought about by agricultural intensification (cuckoos are unusual among our birds in feeding largely on big, hairy caterpillars).

But there may be another, or rather an additional, explanation. Somewhat predictably, given our road-blindness, the possibility that traffication might be involved in the cuckoo's decline has not been investigated. Yet research from the Netherlands showed years ago that the cuckoo is one of the most sensitive species when it comes to traffic noise: numbers start to decline in the presence of road noise as low as 35 dB. Research in the USA has shown that another species of cuckoo, the yellow-billed cuckoo, is also acutely sensitive to traffic noise. A quick check back to Figure 5 (p. 107) shows that even minor roads can produce noise that exceeds 35 dB at a distance of 2 km or more. And looking again at Map 4 (p. 39), which shows all the land falling within 2 km of a road, we see that almost all of England is shaded, whereas large areas of Scotland remain clear of noise. While you have Map 4 open in front of you, take a look again at southern England. In all this area, only Dartmoor, Salisbury Plain and the Brecks of East Anglia remain open as windows of calm – and all of them are hotspots for cuckoos. Indeed, Dartmoor is now just about the only place in the whole of southwestern England that still holds a reasonable population.

Even if road noise is not the driving force of these declines, it might be that traffication has impacted cuckoos by reducing their moth caterpillar food supply. Moths, as we have seen, are profoundly affected by artificial light at night, of which roads and cars are a major source. And if the researchers who found the astonishingly tight relationship between bird numbers and ozone in the USA (p. 121) are right in their suggestion that this might be caused by the toxic effects of air pollution on insects, then yet another plausible link can be drawn between the loss of our cuckoos and the rise of traffication.

The nightingale is another case in point. This subtle songster was once common in hedgerows across southern England, but its numbers have dropped by over 90 per cent since the late 1960s. Theories that have been put forward to explain this collapse include increased grazing by deer of scrubland habitats, the decline of traditional woodland management methods such as coppicing, changes in African wintering grounds and others. No doubt these have all contributed, but the role of traffication surely merits further investigation, if only because nightingales nest (or, rather, nested) in hedgerows, and hedgerows tend to run alongside roads. The same may also be true of another migratory hedgerow species now in desperate trouble, the turtle dove.

These patterns may have other explanations, of course, but surely traffication demands further attention. I am not suggesting that existing theories about our biodiversity loss are wrong, rather that our road-blindness may have led us to miss an additional and potentially even more important cause of the declines that the conservation movement is striving to reverse. It would be a fairly small job, in research terms, to work out whether traffication is at least a plausible explanation for the decline of the cuckoo, nightingale, turtle dove and other threatened species.[51]

The benefits of doing so are clear: if we find that the problems facing our vanishing wildlife have their origins in our well-regulated road system, rather than in the vast expanses of the African Sahel or in the global pandemic of climate change, then they immediately become very much easier to fix. Even if it turns out that traffication is only a relatively small part of the problem, reducing its impacts might still throw many species a lifeline; one of conservation's core paradigms is that reducing one known threat to a species usually enables it to cope better with others. This is exactly the strategy being proposed to mitigate the impacts on wildlife of climate change: if we cannot address the underlying problem of rising temperatures quickly enough to prevent catastrophic losses of biodiversity, then the next best thing we can do for our wild species is to help them to adapt by reducing the other threats they face. Traffication might be one of the lower-hanging fruits in this regard, for reasons that will become clear.

Winds of change

Opposition to the car started as soon as the first vehicle came chugging down the road and, despite the best efforts of the forces of motordom, it has never entirely been extinguished. Now the voices of dissent are rising again. Some of the world's most heavily trafficated countries are already moving to tackle the car's devastating impacts on human health and its contribution to climate change. In the UK, financial incentives are offered to owners of less polluting vehicles, and targets have been set for the complete replacement of petrol by electricity by 2035. Streets have been closed to traffic and the bicycle is enjoying a resurgence in popularity not seen since its late-Victorian heyday. There is a growing interest in, and need for, mindfulness and quietude, and a rising awareness of the restorative properties of natural sights and sounds. Organisations such as Sustrans, The Countryside Charity (CPRE), Campaign for Better Transport, Transport and Environment, Living Streets, Reclaim the Streets, Campaign for National Parks, Mums for Lungs, Quiet Parks International and others, many of them founded in the last few decades in response to a groundswell of unease about the impacts of unfettered motoring, have as one of their main aims a reduction in traffication (though none uses that word). In 2022, the Olympic cycling champion Chris Boardman was appointed head of a new body, Active Travel England, and promised to take back England's streets from motor traffic and return them to cyclists and pedestrians. The tide is slowly turning against the car, in cities at least.

Only the nature conservation lobby, it seems, is absent from the de-traffication party. But in seeking to build on this momentum to the greater benefit of wildlife, they might find themselves standing shoulder-to-shoulder with many new allies (and potential members), pushing together at a door that is already half-open. And there is a further incentive to engage in this arena: traffication is a severe environmental problem, but it offers comparatively cheap and easy solutions. As we shall see, we can reduce many of its impacts without losing the undisputed freedoms that our mechanical brilliance has brought us. Every driver can do something to reduce the impacts of traffication, but they cannot become part of the solution until they know they are part of the problem, and that cannot happen until they are aware that a problem exists to begin with.

Surely the time is right to make this a new direction in the conservation of our country's beleaguered wildlife?

I like to think of Dayton and Lillian Stoner nodding their agreement from the afterlife.

CHAPTER 10

Winners and Losers

One of the many wonderful places I have been lucky enough to work during my career as an environmental scientist is the steppe region of central Kazakhstan, where for nearly 20 years I have been involved in research on a beautiful but perilously endangered bird called the sociable lapwing. Both its common and its scientific name (*Vanellus gregarius*) suggest a degree of aggregation that it actually displays at only a few brief times each year. During migration, sociable lapwings gather together in flocks that sometimes number in the thousands, but during the breeding season they nest in lone pairs or small colonies that are scattered gauze-thin across the vast beauty of the Kazakh steppe. Our research into their ecology and threats therefore requires us to do a lot of driving.

The car we use is the Lada Niva, a small, no-frills four-by-four that is the workhorse of the steppe. What it lacks in comfort the Niva more than makes up for in its ease of repair, a quality you quickly learn to appreciate when you break down two days' walk from the nearest village. I once watched in admiration as my friend Maxim, a native of the steppes, fixed what seemed to be a terminally *kaput* Niva using nothing but a spanner and a flamingo's tail feather.

Villages and small towns in the steppe are scattered almost as far apart as the lapwings, the larger ones connected by raised roads that date back to the Soviet era. Infrequent maintenance, coupled with tarmac-melting heat in summer and irresistible frost-heave in the bitter cold of the steppe winter, have rendered some of these highways almost impassable, little more than a honeycomb of deep potholes that can be frustratingly slow to traverse. Even the slow and steady Stoners might have been exasperated by the rate of progress, and there would be no roadkill to keep them entertained along the way as nothing can travel fast enough to pose a risk to even the slowest animals. Not surprisingly, local people try to avoid the worst of the potholes whenever possible, and where the ground is dry enough they create their own impromptu dirt tracks that cut out across the open steppe. One driver pioneers a way through, leaving no more than two parallel tracks of silver where the pale undersides of wormwood leaves are bent upwards by the vehicle's passage. A second driver sees these tell-tale signs and follows, a third takes the same route and within a few days a fresh new track appears. Cars can travel much faster along these dirt tracks than on the cratered tarmac, and despite a vehicle flow that might number in single digits each week, it is here that we start to encounter roadkill.

In many places, though, the surrounding steppe is too waterlogged for off-road tracks and traffic is forced back onto the sourdough tarmac. What relieves the frustration of the potholes is the wildlife that these roads attract. These are not invasive species, hitching a ride along roads as they conquer new territory; they are natural denizens of the steppe. The black lark, which like the sociable lapwing breeds only in the steppe grasslands of Central Asia, reaches perhaps its highest numbers along the edges of these broken-up highways. What attracts them is unclear, but the roads must provide a good source of the grit they need to break down the husks of seeds in their gizzards, and perhaps the potholes accumulate wind-blown seeds or grain spilled from the occasional overloaded farm truck that jolts through. The roads are often raised well above the surrounding steppe on bunds, so they may provide a good lookout against predators or act as a springboard for the male black lark's wonderful butterfly-like display flights. Whatever the reason, the coal-black males are a constant companion as you edge cautiously down these pitted highways, running around with their peculiar shuffling gait (the smudge-brown females, on the other hand, are far more elusive).

Black larks are not the only beneficiaries of these steppe roads. Wheatears and other birds also use the roads and their raised banks for feeding. Where roads cross streams or rivers they are raised on bridges that provide nesting sites for colonies of lesser kestrels, and culverts running under the road are home to breeding sparrows and swallows. Power lines run parallel to the roads, providing perches for steppe eagles, long-legged buzzards and other birds of prey, including snowy owls in winter. Thin lines of roadside trees planted as wind-breaks are home to nesting colonies of red-footed falcons. Drainage ditches cut alongside the road create small wetlands ideal for nesting marsh sandpipers, pallid harriers and bluethroats. Just by being different from the miles of open steppe grassland around them, these roads offer a range of resources that are otherwise rare in the landscape and many species are able to profit from them. The near-absence of traffic no doubt adds to their attractiveness.

This attraction to roads is certainly not unique to the steppe grasslands of Central Asia. Almost as soon as it was recognised that the impacts of roads could be felt far beyond the tarmac itself came the realisation that different species react in different ways to this new arrival in the landscape. The five wader species studied by Jan Veen in his pioneering research in the Netherlands in 1973 all showed negative responses, their numbers falling with increasing proximity to roads. However, subsequent studies have shown that other species are relatively unaffected by roads and some even respond positively to their presence. One of the early studies to examine how roads influence the number of animals living nearby was undertaken in Illinois, just a few years after Veen's work in the Netherlands. This looked at how the numbers of two bird species, the red-winged blackbird and the horned lark (better known in Britain as the shore lark), changed with proximity to a road. Like so many of the other road-shy species we have encountered in previous chapters, and like all of Jan Veen's grassland waders in the Netherlands, numbers of breeding horned larks were reduced near roads and gradually rose with increasing distance from the verge. The extreme road-shyness of horned larks has been confirmed by a number of subsequent studies. However, red-winged blackbirds showed exactly the opposite pattern – they were found almost exclusively close to roads, and their numbers dropped sharply as the researchers looked further afield. Like the black larks in the Kazakh steppes, red-winged blackbirds clearly benefit from some resource offered by roads, probably in this case roadside tree-lines for nesting, and they are not deterred by the traffic. Their roadkill rate is presumably higher than that of the horned larks, but the advantages of living near the road must outweigh the risks.

Some species, then, benefit from the presence of roads, whereas others lose out. And as a result, strange things start to happen to our wildlife communities.

The rise of wrens and robins

The UK's bird community is hideously unbalanced. Between them, our seven most abundant species account for half of all the country's individual birds, and our two commonest species, the wren and robin, between them make up almost a quarter (see Figure 9). Animal communities in pristine environments are usually much more diverse, their constituent animals or plants distributed more evenly between species. The overwhelming domination of our bird community by a tiny handful of species is a clear indication that something has gone seriously wrong in our countryside.

Our 11 most abundant birds, those with populations of more than 2 million breeding pairs, are, in descending order of population size, wren,

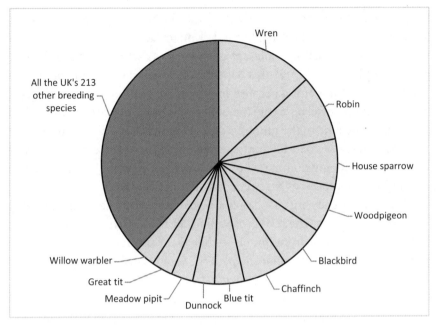

Figure 9. Britain's breeding bird population is dominated by a handful of common species. The circle represents all the breeding birds in the UK (around 80 million pairs), and the pale segments show the 11 species that have populations over 2 million pairs. The populations of all the other 213 species combined are shown by the darker segment. Just 5 per cent of our breeding species account for over 60 per cent of all the individual birds in the UK. Data from the 2020 State of the UK's Birds report (Burns *et al.* 2020).

robin, house sparrow, woodpigeon, blackbird, chaffinch, blue tit, dunnock, meadow pipit, great tit and willow warbler. Two of these, meadow pipit and willow warbler, owe their place in this exalted ranking to their high abundance in upland habitats, usually in areas with rather low densities of roads, and in fact their populations in the heavily trafficated lowlands are falling. But the other nine species are all common in heavily trafficated lowland regions.

Of course there are plenty of possible explanations for this pattern, but an interesting question leaps out at us: is it possible that the UK's bird community has become so utterly dominated by a handful of abundant species *because* of traffication?

This is not an implausible suggestion, and the process could occur in two ways. First, species that are sensitive to the impacts of roads will be greatly reduced in numbers and distribution across much of our heavily trafficated country, leaving a few road-tolerant species to dominate our bird populations. The road-tolerant species will naturally float to the top of the league table of abundance simply by not declining in numbers near roads while others around them do. Second, it might be that our most common species are more than just tolerant of roads, and that they actually benefit from them in some way. If this happens, they will increase in abundance near roads and therefore, because pretty much everywhere is near roads, they will pull themselves to the top of the table. If either of these two things happens – a decline in the numbers of road-sensitive species or an increase in the numbers of road-tolerant species – then our bird community will become increasingly dominated by a few species. In fact, there is evidence that both these processes are occurring simultaneously.

We have already met Dr Sophia Cooke, one of Britain's inexplicably few roads ecologists. The results of Sophia's PhD research confirmed earlier patterns such as those from Illinois – while many species show an avoidance of roads, some show the opposite pattern. Sophia tried to explain why this might be by comparing the characteristics of road-shy species with those of road-tolerant species. One of the important patterns she found, which confirm similar findings from a number of other studies, is that migratory birds are generally much more sensitive to roads than species that live here year round. A quick look back at the list of Britain's 11 most abundant species finds that only one of them, the willow warbler down in 11th place (and falling fast), is a long-distance migrant.

But Sophia also detected another important pattern, one that suggests that traffication has fundamentally altered the structure of bird communities across the whole of our countryside. What she found was that our scarcer species – those with a relatively small overall population in the UK – reach their greatest abundance in less trafficated areas, whereas species with high populations at the national level show the opposite pattern: they reach their highest numbers in areas with higher exposure to road traffic. Species with intermediate population sizes tended to show no particular association with roads, either positive or negative.

The first of these findings is exactly what we might expect to see for road-sensitive species: their numbers are greatly reduced near roads, by traffic noise or roadkill or whatever, making them scarce species in a country with a dense network of roads, and their highest numbers persist in areas with lower densities of roads. This reduction in numbers near roads is the road-effect zone. Their numbers in the few remaining undisturbed areas may be little different to what they were in pre-traffication days, but the loss of birds within hundreds or even thousands of metres of our dense road network means that their overall populations across the UK will inevitably be low. As always, we need to be careful not to interpret broad patterns like this as evidence of cause and effect, but if Sophia's results do not prove that traffication has reduced the populations of many species at a national level, they are at least entirely consistent with that possibility.

The second pattern requires a little more thought. What Sophia found was not simply that our most abundant species are unaffected by roads; if that were the case, then their numbers would be no higher near roads than elsewhere. Instead what she found was that many of our commonest species reach their highest numbers in heavily trafficated areas, and that as the influence of roads falls, so too do their numbers. Our most abundant birds, therefore, show exactly the opposite pattern to that of our scarcer species; just like the red-winged blackbirds in Illinois, they have an inverse road-effect zone.[52] Sophia's analyses showed that of our eleven nationally most abundant species, the nine that are found primarily in lowland habitats all reach their greatest numbers near roads, whereas the willow warbler and meadow pipit, both species that owe their high national abundance to populations in less trafficated upland areas, show a clear negative association with traffication. It is a very neat pattern, and the more I think about it, the more compelling it seems to me.

Quite why some species should reach their highest numbers in the presence of roads, and thereby thrive in a heavily trafficated land, remains unclear. An obvious explanation might be that roads tend to run through certain lowland habitats that just happen to be favoured by these very common species. It might also be that our commonest species owe their abundance to their ability to exploit the plentiful food that householders put out for birds in their gardens; because houses are usually near roads it seems plausible that these garden-seeking species might be most abundant in trafficated areas. However, Sophia factored out such effects by taking account of human population density, habitat and geographical location in her analyses: our commonest species are found in their greatest numbers near roads whether people and their gardens are present or not, and in any habitat, across the whole country.

This suggests that, just as in the steppes of Kazakhstan, roads themselves can offer certain resources that are scarce elsewhere, and those species capable of exploiting them will inevitably become commoner in their vicinity. Such resources might include grit for grinding down seeds in the gizzard, as has been suggested for the house sparrow, or roadside features such as hedgerows, tree lines, grassy verges or fences, which offer birds places to nest, feed and perch. The attractiveness of such areas might be increased by the generally low numbers of predators that occur near roads, which might compensate for the increased risk of roadkill. And since road-shy species have been eradicated from heavily trafficated landscapes, the road-tolerant species benefit from unrestricted access to all the resources they would previously have had to compete for. A woodland purged of its road-sensitive species will support higher populations of road-tolerant species as a result.

These ecological patterns only go so far towards explaining the process, because they raise a further question: how is it that some species are able to exploit the resources offered by roads without suffering all the effects of traffication that other species find so intolerable? This is a question that has long exercised the minds of some of the world's leading road ecologists, such as Canadian scientists Lenore Fahrig and Trina Rytwinski.

In 2009, Fahrig and Rytwinski published an important review that pulled together data from nearly 80 studies looking at the responses to road traffic of over 130 species spanning a wide range of animal groups. They found evidence of negative impacts in 60 per cent of cases, and in 30 per cent of cases could discern no effect either way, but in 10 per cent of cases they detected evidence of a positive impact. (In a later assessment, based on new data, they raised the proportion of species showing positive impacts to around 20 per cent.) Just as Sophia Cooke was to do later with her data on Britain's birds, they compared the characteristics of species showing negative impacts of roads with those that seem to benefit from their presence. In the case of birds and mammals, those most likely to gain in trafficated landscapes were predicted by Fahrig and Rytwinski's analyses to be smaller species with high reproductive rates that defend small territories and therefore tend not to move very far. These are characteristics typical of non-migratory songbirds, so it may therefore be no coincidence that of the eleven commonest species in the UK, ten are songbirds (only the woodpigeon is not) and ten are not migratory (only the willow warbler is). Small territories and high reproductive rates may offer such species some protection against roadkill: they are less likely to cross roads than species that need to move around to defend large territories, and their high output of eggs and chicks might allow them quickly to replace those that fall victim to traffic. A quick look down our list of 11 commonest species identifies several that are unlikely to be frequent road-crossers, not least the weakly flying, stubby-winged wren up there in top place.

How these road-tolerant species manage to overcome the other ill effects of roads, such as traffic noise, is less clear, but overcome them they clearly can. Experimental studies in the field of road ecology are rather rare, at least outside the laboratory, because researchers generally have little influence over traffic flow. But just occasionally scientists are able to take advantage of experiment-like circumstances. In the early 1990s, some German ornithologists learned of a new four-lane autobahn that was to be constructed between Steinau and Schlüchtern in the state of Hessen. Seizing their opportunity, they put out a huge number of nest boxes for breeding birds, both along the new highway as it was being built and at sites further away (to act as controls). They monitored the birds nesting in them, mostly blue tits and great tits (both of them in Britain's top-ten list of commonest species), for three years before the road opened to traffic and for two years afterwards. Each year they measured the number of eggs laid in each nest box and carefully weighed the chicks as they developed. And when they came to examine their data at the end of the five-year study, they could find no discernible impact of the road on breeding blue tits and great tits. Nest boxes placed near the road were no less likely to be occupied by birds than those further away. Birds nesting near the road were just as productive as those nesting elsewhere, both in terms of clutch size and the health of the chicks after hatching, and they were just as successful after the road opened to traffic as they were before. Several subsequent studies of these two species have yielded exactly the same results – blue tits and great tits seem to be utterly unaffected by any aspect of road traffic. Indeed, one study even suggested that blue tits actually prefer nesting in boxes with artificial traffic noise being played inside them to boxes without.

Studies of another of our most abundant birds, the house sparrow, have yielded similar results: urban sparrow chicks show no adverse effects of traffic noise, either in the levels of stress hormones in their blood or their growth rates. Similar results have been found for the starling (which falls just outside our top 11 list of commonest species): constant traffic noise appears to cause the nestlings in this urban-breeding species no stress at all. The French scientists who carried out the research on urban house sparrow nestlings followed it up with another study that subjected rural sparrows to artificial traffic noise and found that, if anything, they actually benefited. We have no explanation for why this should be.

If some species can thrive in heavily trafficated areas by being somehow innately immune to the effects of noise, others can respond to the same problem by adapting to it. We have seen how birds and other animals can attempt to overcome traffic noise by singing louder and higher. A number of studies have shown that those species that have higher-pitched voices, or which can raise the pitch of their voices in response to traffic noise, tend to survive better in trafficated areas, whereas those with lower-frequency songs are usually evicted by the din of passing vehicles. The wren and robin, which between them account for nearly a quarter of all the UK's individual birds, might be particularly suited to roadside living in this regard. Despite its diminutive size, the wren has a deafening song that will carry well even in the noisiest environments, and the robin is known to be able to change its voice in the presence of traffic (indeed, it was among the very first species for which this ability was demonstrated).

This is all rather speculative, and we still do not really understand why some species win and others lose near roads. But these results do at least suggest that some species owe their ability to thrive in heavily trafficated areas partly to their innate insensitivity to road noise or to their ability to overcome it, and partly by having a lifecycle that makes them relatively resilient to the toll of roadkill.

Whatever the reasons, the combined impact of these processes is to change, very profoundly, the structure of our bird communities across the whole country. In a land where the car is king, our bird communities will inevitably become increasingly dominated by a small number of road-resilient super-adaptors. Species that once occurred together have now been pulled apart by traffication, some pushed away from roads and others gravitating towards them. As a result, many places now support fewer species than they used to, even if the overall number of species breeding across the country as a whole remains largely unchanged. Our bird communities have become less diverse through the loss of some species and the unnatural abundance of others. This pattern is apparent globally: a study that pulled together data from many studies around the world found that the number of birds present in areas affected by roads may not always be lower than that in road-free areas, but the species composition of the two communities is consistently different, each less than the natural sum of their parts.

A further effect of these patterns is that communities not only become less species-rich; they also become more similar to each other: bird communities near roads now comprise the same few common species of traffic-tolerant birds whether you are in Sussex or Sutherland, in Cornwall or Caithness. Regional variation in bird communities has been eroded. This is a commonly observed phenomenon in areas that have been severely degraded and has been given its own technical name: *biotic homogenisation*. What it means is that our wildlife communities have become simpler and less varied. Roads have fundamentally altered bird communities across our entire landscape. The wrens and robins may have won out from our love of the car, but it is a somewhat hollow victory when compared with what we have lost. Traffication has robbed our landscape of much of its natural diversity in the same way that it has eroded our cultural diversity.

Benefits of life on the road

Whether similar patterns occur in other groups of animals is unclear: complex statistical analyses such as those undertaken by Sophia Cooke require huge amounts of data, particularly if the effects of other factors such as habitat and human habitation are to be accounted for, and at the moment we only have such large datasets for birds. However, there is plenty of evidence that small mammals such as rodents often occur in particularly high numbers on roadside verges, and many seem to show the same pattern as the wrens and robins: their numbers peak at the roadside and gradually fall as the road's influence is reduced by distance. This is entirely in line with the prediction made by Lenore Fahrig and Trina Rytwinski that the species best able to thrive near roads are those with small territories and high reproductive rates. Few mammals can reproduce as quickly as rodents, and we have seen already that most of them are infrequent road-crossers.

For animals that can tolerate the roar and pollution of passing vehicles, and whose lifestyle does not require them to cross the tarmac on a regular basis, life by the roadside might offer some sizeable advantages. For a start, numbers of larger, slower-breeding and more wide-ranging predators are generally lower near roads, again in line with Fahrig and Rytwinski's predictions. For small animals living by roadsides, this may be more than enough compensation for the increased risk of roadkill. For example, freshwater turtles in North America often choose to lay their eggs in roadside habitats despite the increased risk of roadkill because their eggs have a better chance of survival owing to the absence of predators. At a population level, the benefits of higher egg survival outweigh the costs of the higher risk of roadkill.

Just as many non-native species use roadsides as routes of invasion, so too can native species use them as convenient corridors of movement through the landscape. Research from Germany, for instance, has shown that the hazel dormouse is rather tougher than its impossibly cute appearance might suggest, as animals regularly travel up and down lines of shrubs bordering busy motorways. Roads can open up foraging space for aerial species such as bats and butterflies in heavily forested areas; in Poland, for example, low-traffic roads are more attractive to feeding bats than forest tracks through dense woodland. Road lights may attract and concentrate insects, helping some species of bats to feed.

Even the red stain of roadkill might bring some benefits to those able to profit. Scavenging birds such as buzzards, ravens and red kites find a reliable source of food on our roads. As their populations continue to bounce back after a long era of persecution, roadkill might become an increasingly important food resource, even if exploiting it carries a risk. In the world's roadkill capital of Tasmania, where many of the native mammalian scavengers have been wiped out, forest ravens now feed almost exclusively on the super-abundance of flattened fauna. Studies in Patagonia have shown that smaller scavenging birds of prey such as caracaras are extremely adept at finding roadkill and are generally found close to roads, whereas larger and less manoeuvrable scavengers, such as the gigantic Andean condor, stay well away.

Roads might bring benefits to species passing over them too. Migratory species are particularly sensitive to road noise during the breeding season, but they may be able to use roads as navigation aids during their migration. The best example of this comes from some remarkable studies of homing pigeons. These birds are perfect subjects for studying the mechanics of bird navigation, because researchers know that if they fit a small tracking device to the bird before releasing it, there is a very high chance that they will be able to retrieve it later in the loft and download the precious data it has collected. Professor Tim Guilford of the University of Oxford, who led the pigeon-tracking research team, summarised some of their findings: 'One pigeon flies along the road to the first roundabout, takes the third exit, goes along the dual carriageway to the next roundabout, then leaves the road and goes cross-country.' Roads were not the only linear features used by the pigeons – they followed rivers and railway lines too – but they make up most of the straight lines birds see as they fly over the landscape. By following roads the pigeons may not be taking the most direct route back to the safety and food of their loft, but they accept the cost of a slightly longer journey in return for the very obvious advantage of not getting lost.

Whether wild birds use roads in the same way is unclear, but they certainly use other linear features in the landscape, such as rivers, so it seems entirely possible that roads play a part in their navigation during migration. The Levant sparrowhawk, a scarce species that migrates between its breeding grounds in south-eastern Europe and south-western Asia and its wintering grounds in east Africa, uses roads in a slightly different way, by taking advantage of the warm air that rises from road surfaces to gain height during its migration. This saves it up to half the flight energy it would otherwise need to expend, and because roads heat up faster than other surfaces, these rising columns of air (thermals) start forming sooner in the day and so allow the sparrowhawks to get an early start.

Slivers of green

Of all the wildlife habitats that have diminished in area in the UK over the last few decades, few have lost more ground than our flower-rich semi-natural grasslands. Since 1945, over 95 per cent of our meadows have been lost, most converted to intensive silage production or built over entirely. But strips of grassland often survive along roadsides, protected from further development or cultivation by the hurtling traffic. When that great doyen of British conservation Derek Ratcliffe wrote his hugely influential *Nature Conservation Review* in 1977, he estimated that Britain had over 2,000 sq. km of roadside verges, the combined size of an average county, many of them supporting plants and insect species of national significance. Ratcliffe saw that these represented a new type of habitat in Britain, because although many road verges are of great antiquity, the evolution of roadside verges into the distinctive habitat they are today followed the arrival of cars. Before cars arrived roadsides tended to be heavily grazed, and hence heavily fertilised, by livestock, but now they give grazing-intolerant species an opportunity to flourish. The construction of roads often strips away topsoil and brings underlying geology such as chalk to the surface, allowing distinctive communities of plants and insects to evolve and flourish, if left alone for long enough to do so. The older the roadside verge, the more grassland specialist plants it tends to support.

With growing recognition of the distinctive and important wildlife communities that even short lengths of roadside verge can support, there is now a great deal of interest in trying to claw back from the car some sort of compensation for our traffic-bludgeoned wildlife by managing and protecting the best of these verges as reserves. A number of counties, particularly Norfolk, Suffolk, Bedfordshire, Kent and Lincolnshire, have led the way in designating and managing Roadside Nature Reserves (RNRs). Inevitably, given the toll of roadkill and noise pollution, interest tends to focus on the plants and invertebrates that these can support. Because verges are mown rather than grazed, managers have the opportunity to time their intervention to optimise the gains for wildlife. Cut them too frequently and nothing survives long enough to set seed. Cut them too infrequently and grasses and scrub will start to take over. Leave the cuttings on the ground and soil nutrient levels rise, allowing a few species of grass to crowd everything else out.

But get the management just right and carpets of wildflowers will flourish along roadsides, providing food for a host of insect pollinators – and perhaps for millions of car-bound commuters their only daily contact with wild nature. Among the common species will hide a number of much scarcer ones, with such evocative names as green-flowered helleborine, lady orchid, lizard orchid and crested cow-wheat. Close to half of all Britain's flowering plants can be found along roadsides. Scarce butterflies can do well here too: Adonis and chalkhill blues, grizzled and silver-spotted skippers. Slow worms, lizards and other reptiles bask in the more open patches.

These narrow strips of grassland represent perhaps the greatest direct contribution to wildlife conservation of our road network. However, efforts to improve roadside verges for wildlife may not benefit all species equally. The flush of green attracts rodents and other small mammals, and in turn the predators that feed on them, which are then vulnerable to being hit by cars. This explains the high death rate of barn owls along roadsides. For this reason, RNR managers often attempt to zone their little reserves, creating a strip of very short vegetation immediately next to the road to deter small mammals and their predators, then gradually increasing the height and complexity of vegetation further from the edge of the tarmac.

Just as with birds and mammals, there are winners and losers in the plant communities of roadsides. Our profligate use of salt on roads harms many species, but offers an opportunity for others. Salt-loving species (halophytes) of coastal habitats such as reflexed saltmarsh-grass, sea plantain, Danish scurvygrass and lesser sea-spurrey have been able to march far inland along ready-salted roadsides. Plants native to Atlantic and Mediterranean coastlines, such as buck's-horn plantain, have spread along highways into the heart of the continent.[53] As with the invasive non-native plants we encountered in Chapter 5, their spread has been speeded by

St George's Flower Bank, a local nature reserve bordering the busy A369 road in south west England. Well-managed sites like this can support important populations of wildflowers and insects and offer harassed commuters what might be their only daily contact with nature. *(Giles Morris)*

vehicle movement. Some plants benefit from exhaust fumes, fertilised by an invisible fog of nitrogen gases from the exhausts of passing cars. Others can quickly adapt to toxins; one of the earliest studies of roadside plants, undertaken back in the days of leaded petrol, found that a grass called red fescue rapidly evolves a tolerance of high concentrations of lead in roadside habitats. A later study found that road-adjacent populations of another common grass, sweet vernal grass, not only develop an increased tolerance to road salt but also adapt in other ways to this new environment, evolving within the space of a few decades to have a different flowering date and even a different shape from populations in less saline environments.

Some plants can adapt to roadside living whereas others cannot, with the result that, just as with bird communities, plant communities near roads start to simplify, a few nitrogen-loving grasses tending to dominate. This is why roadside verges need to be cut regularly to maintain their floral interest, and the cuttings removed, otherwise the less tolerant species would quickly be choked out. On trees in London, nitrogen-tolerant lichens benefit at the expense of sensitive species, spreading to occupy the gaps left by the losers. The effects of traffic on heathland plant communities may be apparent up to 200 m from the roadside. Research in Germany has yielded strong evidence that the fumes from road traffic alter plant communities, because the impacts on forest floor plants could be seen for only 80 m upwind of the road, but for three times that distance downwind of it.

There are, then, a few winners in a heavily trafficated world. Some of our commoner species have benefited from roads; or, rather, it might be more correct to say that those that are able to take advantage of roads have *become* our commoner species. Wrens and robins have gained as other species have lost, and our bird communities have become distorted and simplified as a result. Roadside nature reserves prove that, with adequate management, rich wildlife and busy roads can live side by side. But they are usually only a few metres wide, often narrower than the road itself, whereas the road-effect zone for sensitive species might extend a kilometre or more. These thin slivers of grass and flowers provide some solace for the wildlife we have lost to traffication, but they are not compensation.

If only we could retain the convenience of our cars, together with the few genuine environmental benefits that roads bring, and at the same time reduce or even eliminate the devastating impacts of roadkill, fragmentation, soundscape pollution and all the other terrible costs that our driving imposes on the planet's fragile ecosystems.

Well, the good news is that we can.

CHAPTER 11

Five Reasons for Hope

O ur seemingly incurable dependence on road travel could easily engender a rather glass-half-empty view of the future. We live in a society designed around the car, making it very hard for us to wean ourselves off it. Predictions of the future suggest that all the elements of traffication will continue to increase: based on current trends, the coming years will see us drive further, and faster, in more vehicles, around a longer network of wider roads. Our car addiction appears to be beyond any hope of recovery, dooming us by laziness, laissez-faire and vested interests to further decades of appalling damage to our health, our environment and our wildlife.

This is a bleak view of the future but it is not an irrational one, because it is simply an extension of the past. Our response to congestion has only ever been to build more roads and to widen existing ones, and never to try to reduce the problem at source. But increasing the width and length of our roads does not alleviate overcrowding; instead it actually generates new waves of driving that quickly eat up the available capacity, a phenomenon known as *induced traffic*. The American polymath Lewis Mumford, whose many interests included urban planning, recognised this as long ago as 1955, when he memorably likened the building of more roads to ease congestion to an overweight man loosening his belt to prevent obesity. Build more roads, and people simply drive more. There's nothing that motorists want to fill more than an empty road. So inevitable and so predictable is the spiralling effect of induced traffic that it has been called 'the fundamental law of highway congestion'. Until the COVID-19 lockdowns of 2020 and 2021, our driving has only ever increased, and our only response to the problems it creates has been to permit, and thereby unintentionally to encourage, even more of it.

The arrival, or rather the return, of electric vehicles (EVs) is seen by many as a magic bullet that will wipe away the environmental sins of road travel in one fell swoop. Air quality, particularly in cities, should improve with the switch to EVs, especially if the electricity used to recharge their batteries is not generated by freeing energy that was locked away in the age of dinosaurs. EVs are efficient, converting a much higher proportion of potential energy to movement than combustion engines can. They have many fewer moving parts, maybe just 20 or so, compared to 1,000 or more in combustion engines. They therefore require fewer replacements, use less engine oil and last far longer. The jury is still out on how much EVs might contribute to tackling global warming, although future improvements in vehicle technology and renewable electricity generation are likely to bring the lifetime emissions of EVs down well below those of combustion alternatives. EVs should certainly help to reduce a few of the environmental impacts of traffication.

But traffic noise will not fall much, if at all. Until around 1970, most of the noise produced by a moving car came from its engine, but as technology has improved, making engines and exhausts quieter and more efficient, so the causes of noise have changed. For most vehicles travelling over 30 mph or so, the dominant source of noise is no longer the engine but the interaction between the tyres and road surface (rolling noise) and the air flow over the vehicle (aerodynamic noise). EVs travelling above 30 mph are therefore almost as noisy as petrol-fuelled vehicles.[54] There will be no fall in light pollution, and EVs will not reduce the amount of road salt we use.

Other problems might actually increase with the return of EVs. Pretty much all the damage caused by traffic – to the environment, to wildlife and to our own health – increases exponentially with vehicle speed, and EVs can be *fast*. Of the 130 or so cars shown in Figure 2 (p. 32), only one is an EV and it is very easy to find on the graph – it's the one with the highest speed of any vehicle, up there in the top right corner. Car manufacturers are touting EVs as a way to tackle climate change, but at the same time many of them are producing cars that are so overpowered that any potential gains are immediately lost. As Xavier Daffe of the Belgian motoring magazine *Le Moniteur Automobile* has written:

> Why, in a context of environmental transition … are manufacturers still striving to develop electric (or electrified) cars of 400, 500 or 600 hp, even more? … What is this madness, putting on the market electric or rechargeable hybrid machines weighing more than two tons, whose outrageous power is at best used to move this mass, at worst is of no use … in the context of the 30 [km per hour] zones that are becoming more and more widespread?

Even more modestly powered EVs might travel faster on average over a given distance than petrol vehicles because the powerful torque of electric motors gives them high rates of acceleration, allowing them to reach the driver's chosen speed almost instantaneously. Roadkill rates may therefore rise, because speed and sudden death walk hand in hand. Being heavier than petrol-powered cars, and having greater acceleration, EVs will create more particulate pollution from tyre and brake wear and more toxic dust from the abrasion of the road surface. These are now the dominant sources of air pollution from road transport, responsible for thousands of deaths every year. Half a tonne of battery weight can result in particulate pollution from tyre wear that is hundreds of times higher than the exhaust pipe emissions of a combustion car. Exhaust gases will fall, but particulate pollution is likely to rise.

Our switch from combustion to battery will also bring with it some wholly new environmental problems. EVs contain several times more metal than combustion vehicles of equivalent size, and demand for copper, nickel, cobalt and manganese will rise. Battery manufacture is very energy intensive and requires the use of rare-earth elements and scarce metals such as lithium and cobalt, the sourcing and extraction of which raise many serious environmental, social and geopolitical concerns.

Most worrying, as far as our wildlife is concerned, is that EVs will encourage us all to drive even more than we do now. Once we have swallowed the huge cost of buying an EV in the first place, our driving suddenly becomes more affordable, particularly if we are powering our new vehicle from solar panels on our roof. Drivers of EVs may be misled into thinking that their motoring is now environmentally guilt-free, and do more of it as a result; they may even abandon their bicycles and public transport in favour of their new battery-powered cars. Information on how people might change their driving habits in the coming EV age is hard to come by, as most countries do not record separate mileage data for vehicles using different power sources. It is also difficult to interpret, because early adopters of EVs are unlikely to be representative of drivers as a whole. But data from the Australian Bureau of Statistics show that EVs are driven on average 600 km further each year than petrol-fuelled cars. Very similar findings have been put out by the car manufacturer Nissan, again suggesting that owners of EVs drive around 600 km further each year than those who have not yet made the switch. This might not sound like much of an increase, but it translates into billions of additional vehicle miles on our roads each year. As battery technology improves, giving EVs longer range and faster recharging, we may see an even steeper rise in car use. Of all the different scenarios explored by the UK Department of Transport when it tried to forecast future growth in vehicle volume on Britain's roads, the one that predicted by far the biggest increase was an all-EV future. Just like expansions to the road network, the switch to EVs will induce new waves of driving. The problems of roadkill, habitat fragmentation, noise and light pollution might therefore all intensify, even if some aspects of air quality improve. Taking all these factors into account, a recent review of the ecological costs and benefits of EVs concluded, rather damningly, that they are 'at best a distraction from the many environmental challenges facing transport'.

Traffication is an environmental problem that cannot be solved by the lithium-ion battery alone. But it is, nevertheless, a problem that can be solved.

'You are traffic'

When it comes to the environment, I am not a natural optimist; a career spent studying and writing about habitat destruction and extinction tends to erode your hopes for the future of life on Earth. I have spent years trying to save perilously threatened species that I now fear have little chance of long-term survival; one or two of them may even predecease me.

But in the case of traffication, I feel rather hopeful that things could improve in the next few decades. There are five reasons for this uncharacteristic optimism.

The first is that we all contribute to the problem, and we all suffer its effects; as a now famous satnav advert insightfully pointed out in 2010, 'You are not stuck in traffic. You are traffic.' This might not at first sound like grounds for optimism, but if we are all part of the problem, then we can all be part of the solution. Of course the same is true of other environmental problems – each of us can do plenty of things to reduce our contributions to global threats such as climate change – but the difference with traffication is that we are all more or less equal contributors. Some people obviously drive more than others, but for no other global environmental problem is the potential to fix things spread quite so evenly among us. We do not need to wait for politicians to act before the problem can be addressed, nor are we under the yoke of decisions taken in big-business boardrooms. We can act now. What is more, many of the things we can do now will have an immediate impact. If we were to stop pumping greenhouse gases into the atmosphere tomorrow, it would take hundreds or perhaps even thousands of years for our climate to recover to its pre-industrial condition. But if we were to stop driving tomorrow, most of the impacts of traffication would instantly disappear. Traffication is a problem that can be substantially reduced by the individual actions of concerned people, and there are many things you can do that will immediately lessen the impacts of your driving on the environment. Perhaps all that is stopping people from doing something about it is that they aren't aware there is a problem to start with, and that is fairly easy to put right.

Conservationists are fond of describing solutions to environmental problems as being either 'bottom-up', in other words embraced and enacted spontaneously and willingly by large numbers of people, or 'top-down', whereby people are steered or even coerced by law into doing things they might not otherwise choose to do themselves. If my first reason for optimism falls into the former category, then my second is an instance of the latter: it is that our road system and all the vehicles it carries are highly regulated. All our vehicles are registered and tested each year, and government statisticians collect data on where and when we drive, and how fast. We have laws in place to limit vehicle speeds, to control engine noise and to identify and penalise the most polluting vehicles. We have processes to enforce those laws, and we collect the data to know whether or not that enforcement is working. And those data tell us that legislation *works*: for example, the exhaust emissions of particulate pollution, tightly regulated by law, have fallen hugely over time, whereas non-exhaust emissions from tyre and brake wear, which are not currently limited by any legislation, have risen.

This legal framework extends beyond individual vehicles; major extensions to our road system are required by law to pass detailed environmental impact assessments and to implement mitigation measures where necessary. Almost all our roads are in public ownership so nowhere, and nobody, is exempt from these restrictions. We don't need to start from scratch when it comes to developing a regulatory system to combat a serious environmental concern, as we have had to do with the more recently emerged problem of climate change. The policy framework, the legislative processes and the information systems we will need to reduce the worst impacts of our driving are already in place, and have been for decades. We just need to use them better.

Long and often painful experience has shown that solutions to environmental problems will only succeed if they fulfil three criteria: they need to be effective, they need to be affordable and they need to be socially acceptable. My third reason for optimism is that there are plenty of things that can be done to reduce the impacts of roads that meet this trinity of conditions. They do not require people unwillingly to give up their cars, something that would certainly challenge that third requirement of social acceptability. Indeed, many measures can be put in place without the great majority of drivers even realising, let alone caring. It is quite possible you have driven past, under or over one of them today and didn't notice it. We can massively, and immediately, reduce the impacts of our driving on the environment without having to give up all the freedoms the car gives us.

My fourth wellspring of confidence for the future is that efforts to reduce the impacts of traffication will be supported by rapid advances in transport technology and by continued growth in the science of road ecology. We can drive greener by driving smarter, and there are many extraordinary and brilliant ideas in air for how we could do this.

Sophisticated new road surfaces and roadside baffles are being developed that greatly reduce the spread of traffic noise and can absorb microplastic pollution. Some of them even generate solar power at the same time, allowing roads to be heated and thereby removing the need to dump millions of tons of de-icing salt into the environment each year. A company called Solar Roadways has claimed that if its product were to replace all paved surfaces in the US, they would be able to generate three times the country's electricity needs, with clear benefits for the climate. New tyre designs can cut vehicle rolling noise, the biggest contributor to soundscape pollution, by half. Advances in regenerative braking for EVs will not only recapture otherwise lost energy but also reduce brake wear and cut particulate pollution. Trials are underway to develop plant-based plastics for use in tyre composites to reduce microplastic pollution, and calls are growing for a new government-backed test to make it clear how resistant each tyre is to wear and tear, with information labels for buyers. One company has proposed an airless tyre that can be 3D-printed from organic material and is fully biodegradable; others are developing systems to capture tyre wear particles as they are produced and store them for safe disposal later. The search is also on to find less harmful alternatives to bitumen for road-building, and tests using recycled household waste and by-products of the paper industry have yielded promising results. Other power sources currently in development, such as hydrogen or ammonia combustion engines and hydrogen fuel cells, may bring environmental benefits in terms of reduced air pollution.

Communications technology has a part to play too. The UK Transport Vision for 2050 envisages a future in which all the vehicles on the road at any one time cooperate with each other through advanced mobile and satellite technology to optimise traffic flow and so reduce both energy consumption and journey times. Street lighting could also be connected to these networks, allowing it to be dimmed when no vehicles are around and so reduce light pollution at night. The LED streetlights that have all but replaced the old yellow sodium ones can be programmed to emit light in spectra least likely to harm the local wildlife, perhaps changing in brightness or wavelength over the course of the night or the year to take account of the changing vulnerability of different species at key times in their lifecycles.[55] For instance, streetlights running along beaches where sea turtles nest could switch to red for the brief period when eggs hatch, because red light does not disorientate the hatchlings as they make their way to the sea. We are also investing more in natural technology, identifying trees and shrubs that are particularly good at deadening traffic noise or capturing and sequestering the pollutants churned out by passing vehicles.

Alongside all these exciting technological advances, we are learning more about the environmental problems of road traffic, about where, when and why they occur, and how we can fix them. The world is slowly waking up to the fact that the car is a major problem for wildlife and researchers are rising to the challenge (in some countries at least). The database of scientific publications on road ecology that I compiled while researching this book contains 69 items published between 1991 and 2000, 334 published between 2001 and 2010, and a massive 790 published between 2011 and 2020 (embarrassingly few of them written by researchers in the UK).

As we learn more about the impacts of roads on wildlife, we will be able to develop ever more effective and targeted solutions to the problems of roadkill, fragmentation and pollution. The 1990s research boom on the collapse of farmland biodiversity allowed us to design effective conservation solutions for animals and plants striving to exist in fields of industrialised agriculture, one of the world's most hostile environments. We now know how to put in place measures that can double the numbers of birds such as the skylark in modern wheat monocultures at no cost to the farmer. There is no reason why we can't now develop a range of similarly smart, targeted and cost-effective solutions to the problem of traffication.

Things are moving fast. People being born today will be able to look back in later life at early twenty-first-century images of long, parallel lines of stationary, pollution-belching cars, each with its single stressed and unhappy occupant, and slowly shake their heads in disbelief at their parents' folly.

The winds of de-traffication

My fifth, and greatest, source of hope for the future is that society is already drifting slowly towards de-traffication, blown by strengthening winds of concern over human health and climate change. Long gone are the days when the opening of a new motorway was seen as a matter of national pride, presided over by prime ministers with bottles of champagne in front of massed ranks of photographers. New roads are now fractious affairs, guiltily undertaken and strongly opposed, often ending in pitched battles between contractors and well-dug-in protesters. The national mood has changed, road-building hubris replaced by doubt and self-reproach.

There is a little history to this trend. In 1963, two influential reports were published on Britain's transport infrastructure, one now largely forgotten and the other still infamous. *Traffic in Towns*, better known as the Buchanan Report, was the first detailed review of the direction of road policy in Britain, focusing, as its title suggests, on urban areas. Professor Colin Buchanan, an Indian-born Scottish town planner, started by recognising the growing problems of urban traffic, pointing to congestion, accidents and, very unusually for the time, environmental damage (though only with regard to the human environment – wildlife didn't get a look-in). Measures such as one-way streets, flyovers, urban freeways and multi-storey car parks would be needed to reduce the problem, Buchanan predicted, but they could not prevent cities from eventually becoming choked by congestion: 'All the indications are that given its head the car would wreck our towns within a decade.' Only in small towns, Buchanan concluded, could normal life co-exist alongside unfettered car use. However, his report recognised that carmageddon was not inevitable. Citing the examples of Venice and a handful of other largely vehicle-free cities, Buchanan saw that urban functioning need not be reliant on the car. These insights led him to develop the concept of the 'environmental area', within which traffic is greatly reduced or excluded altogether and where pedestrians have priority, even as he saw the need to increase vehicle flow elsewhere.[56] Buchanan's view, which ran contrary to contemporary economic thought, was that environmental standards should be absolute, not to be traded away in favour of other considerations. If the modern de-traffication movement needs a birthday, then perhaps the publication of the Buchanan Report in November 1963 is as good as any.

Buchanan's report on how to deal with the rapid growth in road traffic was published just a few months after another report that would only exacerbate the problem. *The Reshaping of British Railways*, more infamously known as the Beeching Report (in fact, the first of two reports by the still deeply reviled Dr Richard Beeching), led directly to the closure of thousands of miles of railway line and around half the country's stations. These cuts were claimed by Beeching (and his no less controversial political paymaster Ernest Marples) to be necessary because of the growth of road travel, but they were much lamented at the time and continue to be so today; some of the abandoned lines, indeed, are now being reinstated at huge cost.

In the same year, therefore, the British government was presented with one report calling attention to the problems of rising road traffic, and another arguing for drastic cuts to its most viable alternative. In the end it was Beeching's recommendations that were enacted with greatest zeal, setting us firmly down the road to traffication. But now, 60 years later, it is Buchanan's ideas on the primacy of a healthy human environment that are in the ascendant.

We have not ended up with this heavily trafficated world because the car is the best solution to our transport needs. We have been dragged here, more or less willingly, by flawed policies and vested interests that have locked the car so deeply into our society that escaping it now seems simultaneously essential and unthinkable. Propelled by powerful manufacturing and petrochemical interests, the car has been unfairly advantaged over other forms of transport from the very start of motoring.

But things are changing. We are starting to realise that the benefits of de-traffication to ourselves and to our wildlife could be enormous. Roads carry huge costs to our physical and mental health that have now become impossible to ignore; in its ability to blind us to its severe health risks long after they became scientifically undeniable, the car is the natural successor to the cigarette.[57] It generates and perpetuates social inequalities and fuels a dependence on oil and other raw materials that forces us into obsequious relationships with distasteful political regimes.[58] Many of the economic arguments used to justify the construction of major new roads are later found to be false promises, the claimed benefits in terms of shorter journey times being negated by induced traffic. And to top it all, roads are a major contributor to climate change, the greatest existential threat we have ever faced.

It may be a little too soon to say that the tide has turned against the car, but we might at least be approaching the high water mark of motoring, what some writers refer to as 'peak car'. As we saw in Chapter 9, the last few years have seen the appearance of many new organisations determined to reduce the harmful impacts of traffication, both by restricting vehicle use and by making alternative forms of transport more accessible and attractive. These organisations have as their starting point different motivations: some are concerned about the aesthetic impacts of roads on our countryside, some about climate change, some about road safety and some about pollution. But all of them are working towards the same goal of de-traffication.

Just as traffic can be induced by building new roads, so it can be un-induced by closing existing ones. Experience from a number of cities has shown that where roads are closed to cars, or when parking space is restricted (a particularly effective strategy to deter car use), there is little or no corresponding increase in congestion in the surrounding areas – the traffic prevented from using the roads simply evaporates. People choose either to travel in other ways or not to travel at all. Research has found that when people are forced to find different ways to travel they often benefit by discovering more efficient routes or modes of transport and may never return to their cars. The Low Traffic Neighbourhoods (LTN) scheme in London uses street furniture to close certain roads to cars: this does not lead to congestion elsewhere; instead research shows that many people readily abandon their cars and take up walking or cycling.[59]

Experiments designed to allow people to see what a de-trafficated world might look like, such as car-free days, are gaining in popularity. In response to COVID-19 lockdowns, some cities began closing streets to cars to create calmer environments to facilitate socially-distanced recreation and exercise. The popularity of these temporary car-free initiatives encouraged people to demand that these spaces remain car-free after restrictions were lifted; during 2020, road traffic dropped by around a fifth compared with the pre-lockdown year of 2019, but cycling enjoyed its biggest annual rise since the 1960s. Over 80 per cent of all the journeys made in Britain, by any mode of transport, are of less than 10 miles. Perhaps all that stops us from getting the bike out more often is that there are too many cars on the roads so we decide to drive instead, a cycle of illogicality we can surely break.

Even the British government, historically (like most governments) a stalwart champion of roads, seems to see the writing on the wall. A Department of Transport report entitled *Decarbonising Transport*, published in 2020, opened with a ministerial summary that envisioned a future in which 'Public transport and active travel will be the natural first choice for our daily activities. We will use our cars less and be able to rely on a convenient, cost-effective and coherent public transport network.'[60]

Traffication is driven by hugely powerful car-making and petrochemical industries that history has shown to be very resistant to any proposed changes that run counter to their economic interests. The clearest example of this is the infamous case of leaded petrol, now thankfully locked away in the filing cabinet of history marked 'How was that ever even allowed?' (filed away, perhaps, next to smoking on aeroplanes). The toxicity of lead was recognised centuries, even millennia, before technicians in General Motors' laboratories in Dayton, Ohio, discovered in 1921 that adding it to gasoline improved an engine's performance. Nevertheless, leaded petrol soon became the default fuel for motor vehicles. It would take nearly

60 years of public health lobbying, and incalculable damage to people and the environment, before the first country would prohibit its use (Japan, in 1980), and exactly a century for it to be phased out altogether (Algeria, the last country to use leaded fuel, finally banned it in 2021). Ridding the world of leaded petrol has prevented over a million early deaths each year, led to a rise in our species' average IQ and – because high levels of lead in the body are associated with delinquency – averted 60 million crimes each year. In economic terms this represents a global saving of $2.4 trillion a year, an astonishing 4 per cent of global GDP.[61]

Traffication is driven by powerful vested interests, but the equally potent forces of public opinion and state policy, backed by new technology, are rising in opposition. For the first time in the history of mass motoring, the proportion of young people learning to drive is falling. Sales of new cars powered solely by fossil fuels will end in the UK in 2030 and sales of hybrids in 2035 (although as we have seen, their replacement by EVs is no panacea for people or wildlife). Recognising the many benefits of smaller, lighter vehicles, the European Public Health Alliance advocates a regulatory framework to ensure that by 2030 all new vehicles should be 'LISA' (Light and Safe), their weight, power, top speed and design all optimised to minimise their environmental and health impacts. They argue that 'The current trend towards ever bigger and heavier SUVs cannot continue.'

Many experts are already seeing beyond the EV to the domination of the autonomous (driverless) vehicle, and further beyond that into all sorts of transport futures in which the car as we know it today has no part to play. They may be writing the car's obituary a little prematurely, but it is probably safe to say that the age of traffication, if we date it from around 1900, has already run well over half its course.

Conservation always works best when it runs parallel to the prevailing zeitgeist, and de-traffication is, in a phrase much loved by environmentalists, a win–win solution for people and wildlife. The only surprise is that the nature conservationists are lagging so far behind in society's rush to curb the worst impacts of the car.

These, then, are my five reasons for hope. As extinction-driving, landscape-splintering, wildlife-slaughtering, soundscape-shattering, pollution-spewing, climate-changing, health-wrecking global environmental catastrophes go, traffication might actually be quite an easy one to fix.

CHAPTER 12

The Road to De-Traffication

There is a sentence in Chapter 11 that is so fundamental to efforts to reduce the impacts of traffication that I will risk repeating it, now with the added emphasis of italics. If you remember just one thing from reading this book, please let it be this: *pretty much all the damage caused by road traffic – to the environment, to wildlife and to our health – increases exponentially with vehicle speed.* The key word here is *exponentially* – a drop in speed of a mere 10 mph might halve some of the problems of traffication, such as road noise and particulate pollution. If we all suddenly started tootling around at 25 mph as Dayton and Lillian Stoner did in the 1920s, many of the environmental problems described in the previous chapters of this book would disappear overnight. This is not likely to happen, of course, but if it did it would not greatly increase the time of the average British car journey, which is just 8 miles (20 minutes at Stoner-speed). In urban areas, journey times would actually fall because slowing traffic improves vehicle flow, more than compensating for lower speeds. Computer models suggest that reducing the urban speed limit from 30 mph to 20 mph would in fact result (if everyone stuck to it) in shorter journey times. The Stoners recognised the importance of speed in the very first study of roadkill ever published, and it remains one of the central pillars of the new science of road ecology.

With a very small number of exceptions (such as Australia's Northern Territories, some of Germany's autobahns and, for some reason, the Isle of Man), the world's roads are at least nominally governed by speed limits. Historically, speed limits were imposed to reduce accident rates. In the UK there were speed limits on the road before there were cars, put in place to protect people from agricultural steam engines travelling at no more than walking pace. More recently, they have also been used to reduce congestion, changing in response to peaks and troughs in vehicle volume. But their greatest benefit, generally ignored by those who argue that limits should be raised, is that by containing vehicle speeds they keep a lid on a wide range of environmental and human health problems. As vehicles travel faster, so they become exponentially noisier, more polluting, more dangerous to crossing wildlife and more damaging to human health. Higher driving speeds are associated with rapid acceleration and hard braking, which massively increase tyre and road wear and hence the production of microplastics and other particulates. In terms of noise pollution, reducing the average speed of traffic from 40 to 30 mph is equivalent to halving the number of vehicles on the road. Researchers examining the impacts of a lowered speed limit in the Swiss city of Lausanne found that the slightly reduced number of road accident casualties, although clearly very welcome, paled into insignificance beside the health benefits to the wider population of reduced noise pollution. In the USA, it has been suggested that reducing speed to bring noise down by just a few decibels would lower the prevalence of hypertension and coronary heart disease with an annual economic benefit of billions of dollars – and that estimate was based on traffic noise data collected over 30 years ago.

Roads carrying fast traffic pose two serious threats to wildlife. The first is that animals crossing them face a high risk of being killed; the second is that fast-moving vehicles and their loud noise deter animals from crossing in the first place, exacerbating the problem of fragmentation. Speed limits, therefore, can be a key tool to mitigate some of the most important environmental impacts of traffication. Lowering vehicle speeds is a much cheaper and more effective way of reducing the impacts of road traffic on wildlife and human health than allowing drivers to go as fast as they like and trying to mop up the resulting damage in other ways. Reducing the speed at which vehicles can travel also means that the cost of damage reduction is borne by the would-be speeder, through fractionally longer journey times, in line with the 'polluter pays' principle. To quote from a report commissioned in 2009 by the UK Noise Association, 'cutting speeds is the most immediate, the most cost-effective and the most equitable way of reducing traffic noise'. New laws now require cars to be fitted with speed limiters, which prevent cars from exceeding the limit of the road they are driving along; currently the law will allow drivers to turn them off, but over time this may change to require them to be permanently activated. Motorbikes, by far the most soundscape-shattering vehicles on the road, might one day be similarly restricted.

Particularly sensitive wildlife areas, national parks for instance, or areas with high densities of large mammals that pose a collision injury risk to drivers, often have speed limits set lower than road conditions might otherwise allow. If drivers comply with the reduced speed limit, roadkill rates inevitably fall and landscape permeability increases. Research from the USA suggests that visitors to national parks are prepared to pay the price of restrictive speed limits in return for the substantial gains in tranquillity they bring.

However, simply lowering the speed limit displayed on signs posted by the side of the road may not be enough to slow drivers down. In a rare example of experimental speed limit manipulation in the field of road ecology, researchers working on some roads in rural Wyoming were given permission to reduce the posted night-time speed limit from 70 mph to 55 mph to see what impacts this might have on the crossing behaviour and roadkill rates of mule deer. Their results were not encouraging; while many drivers slightly dropped their average speed in response to the new limit, the majority still exceeded 55 mph during the critical roadkill hours of darkness and there were no benefits for the deer. The problem, it seemed to the researchers, was that on well-built roads designed to carry traffic travelling at 70 mph, drivers felt safe exceeding the temporary 55 mph limit, particularly since they would have known that the chances of encountering a police patrol car in rural Wyoming at night were pretty low. Clearly, any efforts to reduce speeds, whether permanently or at certain particularly sensitive times of the day or year, cannot rely solely on conventional road signs displaying the speed limit.

But other kinds of road signs might help. The first road signs in Britain were erected for the benefit of cyclists during the great bicycle boom of the 1890s. These 'danger boards', as they were known, warned riders of hazards on the road ahead – sharp bends, steep slopes and the like. As the car nudged the bicycle off the road, motoring organisations such as the AA and the RAC began to erect their own signs (part warning, part advertisement) and campaigned for local authorities to warn drivers of the perils on their roads through clear signage. A wide range of designs emerged, with regional differences in their shape, size and iconography often baffling motorists coming from other areas, but in the 1930s signs were standardised through the international Vienna Convention on Road Signs and Signals into the set of symbols that we recognise today. We pass hundreds of these signs every journey we take, passively absorbing their messages while barely noticing their presence. The fact that we take them so much for granted is a tribute to the smart thinking that underpins their design, for what better qualities could a road sign possibly have than being informative without being distracting?

Signs warning motorists of the danger of wild animals on the road have been used for decades, and very few countries now lack them. The species depicted on them are selected to illustrate the most vulnerable or the largest (and therefore most dangerous to the driver) species on the road ahead. The number of different species depicted has greatly increased over time and in some countries a wide range of signs are deployed. In the Czech Republic,

A sign in Australia warning drivers of wildlife on the road, in this case native kangaroos and emu, and non-native camels. Just visible on the road to the right is a dead emu, suggesting that such signs are not a perfect solution to roadkill. Drivers tend to slow down for a while after passing such signs, particularly if the animal depicted is large enough to pose a collision threat to their health or their vehicle, but soon speed up again.
(Wikimedia Commons)

for instance, where local people are able to propose their own designs, there are warning signs depicting otters, susliks, coypu, dice snakes and even snails. The familiar leaping deer sign in the UK was complemented in 2019 by a new one depicting a hedgehog.

Warning signs might make drivers more aware of animals on the roads, even if they don't always cause them to slow down. A study from the USA found that temporary road signs, accentuated by flashing lights, led to a 50 per cent reduction in roadkill of mule deer during their annual migration; another study, this time from Canada, also found that roadkill rates of large mammals fell significantly after warning signs were introduced. In a national park in South Africa road signs depicting a snake led motorists to change their driving behaviour, resulting in a greatly reduced likelihood that they run over snakes (in fact, plastic ones carefully positioned on the road by the researchers).

The pattern that is emerging from all this research is that drivers react fairly well to temporary warning signs, specific to that time and place, and that we are responsive to new iconography. However, we soon start to ignore

Examples of animal warning road signs from the Czech Republic. There cannot be many countries that warn motorists of snails on the road ahead (and it is not really clear how drivers are supposed to respond). From Tryjanowski *et al.* (2021; *Creative Commons*).

permanent, unchanging signs and seem to have short memories: most of us slow down for a mile or two after passing a wildlife warning sign, but our speed soon creeps back up again to what it was before. There is still much research to be done on how warning signs can be designed and deployed to maximise their roadkill-reducing impact. Road ecologists clearly need to start branching out into the murky minefield of human psychology.

Stitching the landscape back together

The Stoners and their contemporaries saw the solution to the problem of roadkill as keeping vehicles below 35 mph, which they considered to be the wildlife-critical speed. By the 1930s, however, they had to concede that this was unrealistic. Road surfaces were improving, cars were getting faster and the genie of mass motoring was well and truly out of the bottle; by 1940, most vehicles on the road had a top speed in excess of 60 mph (see Figure 2, p. 32). Other solutions to the problem of roadkill were therefore proposed, some of them bordering on the comical. In 1935, it was seriously suggested that a net could be stretched across the front of every car to catch birds unharmed. Every so often, the theory went, the driver would stop, disentangle the birds from the net and release them safely by the side of the highway. As far as I am aware, this concept was never road-tested.

Solutions to the problem of roadkill then started to run in two rather more realistic directions. The first was to try to prevent animals from crossing the road at all, either by erecting fences parallel to the highway or by creating a wide strip of bare gravel or other uninviting habitat on either side to deter any creature thinking of crossing it. There are three inherent problems with this approach. The first is that animals *need* to cross roads because, as we saw in Chapter 5, the effects of fragmentation might be even more severe in the long run than the toll of roadkill. Heavy-duty fences put up alongside roads to stop animals from crossing over might reduce the body count on the tarmac, but if those animals subsequently starve because they can't reach the resources on the other side then it is better to let them take their chances. The second problem is that the fencing has to end somewhere, so preventing animals from crossing one road may simply funnel them towards another equally dangerous crossing elsewhere. The third problem with this approach is that roadside verges can potentially harbour important populations of plants and insects, so deliberately creating strips of the most wildlife-unfriendly habitat possible is clearly a very inelegant solution to the problem.

Instead, a second and altogether more successful direction was taken in reducing roadkill – to provide safe crossings for wildlife and to encourage animals to use them as much as possible. Thus was born the idea of the permeable highway, a solution to both roadkill and fragmentation. Fencing is usually used to prevent animals from wandering out onto the road, but its purpose is not to turn them back to where they came from but to guide them towards safe places they can cross, either below the tarmac through underpasses and culverts or above it on bridges.

The concept of wildlife crossings goes back to the 1950s, but their first major deployments for conservation purposes were in Florida and in Canada in the 1980s. The Florida panther (a subspecies of the cougar) was then in severe danger of being roadkilled to extinction by the upgrading to an interstate highway of Alligator Alley, a road running through the middle of the panther's last stronghold in Big Cypress National Preserve. The solution the planners put in place was a system of wildlife bridges and underpasses designed primarily to protect panthers. There are now more than 60 such structures throughout the range of the panther, and roadkill rates have been massively reduced in areas where they have been deployed (even if they remain unsustainably high elsewhere). Other species benefit too; the panther crossings have been used by bears, deer, skunks, bats, birds and even fish. Thanks to the introduction of these crossings, Alligator Alley could be upgraded to a multi-lane highway without flattening the last of the panthers.

At around the same time, expansions were being made to the busy Trans-Canada Highway where it passes through Banff National Park in the spectacular Canadian Rockies. So many animals were killed on it each year that the road was once nicknamed 'the Meat-Maker'. But not any more. Now it has the highest density, and greatest variety, of wildlife crossing structures of any road in the world – 38 wildlife underpasses and 6 bridges with more still being added – and roadkill rates have plummeted even as traffic volume has risen. These crossing were put in primarily to protect motorists from potentially dangerous collisions with large animals, but the long-term studies of wildlife crossing behaviour that followed their construction, led by biologist Tony Clevenger, have been fundamental in designing and promoting wildlife crossings around the world.

What the Banff research showed was that well-designed crossings can save huge numbers of animals from violent death. Grizzly and black bears, wolves, coyotes, cougars, moose, deer, bighorn sheep, wolverine and lynx have all been caught on camera crossing these structures hundreds of thousands of times, each crossing a potential roadkill prevented. These crossings also promote genetic exchange between populations of animals that would otherwise be divided by the road. Clevenger and his colleagues devised clever methods to snag a few strands of fur from passing bears from which they were able to extract DNA. Genetic analyses showed that black and grizzly bears crossing the road in either direction were regularly mating with animals on the other side, ensuring that the genetic health of their wider populations was maintained despite the bisecting presence of a four-lane highway.

The Banff project has also identified some interesting differences in how species use these crossings. Some animals are shy – it takes them a long time, years in some cases, to learn to feel comfortable using the crossings – whereas others start using them even before construction work has finished. Some species prefer crossings that are high, wide and short, while others like them to be long, narrow and low. Some like subterranean underpasses, and others will only ever use open-topped bridges.

A 'green bridge' in Canada's beautiful Banff National Park, where wildlife crossings such as this one have massively reduced roadkill rates and reconnected populations of animals on either side. Key to their success is the fencing that runs alongside the road, funnelling animals towards and over the bridge and preventing them from wandering into danger. *(Wikimedia Commons)*

Animals ebb and flow over these structures in unpredictable ways; as Tony Clevenger has noted, crossings are fixed and permanent features in the landscape, but the ecological processes that they support are highly dynamic. Undetected by the automatic cameras that record the passage of larger animals, a host of smaller creatures cross and recross these structures in their droves: snails, spiders, salamanders and centipedes, bats and beetles. With every large animal that lumbers across, a small ecosystem crosses with it. Invertebrates, the spores of fungi, the eggs of fish and the seeds of wild plants are matted into the animals' fur or trapped between the pads of their paws. The copious fruit eaten by bears is carried over roads in their guts to be deposited (in the woods, naturally) as seeds on the other side.

The road cuts a gaping wound through Banff's fragile natural environment but the crossings act as sutures, stitching the wildlife communities on each side back together and helping them to heal.

The permeable road

The concept of permeable highways has taken off around the world, and a dazzling array of different types of wildlife crossing has been developed and deployed. A whole lexicon has evolved around these structures, which are also known as *ecoducts* or, if they run over the road, *green bridges*. The largest (over 100 m wide) are often termed *landscape bridges*, followed by the smaller *wildlife overpasses*. There are also *multi-use overpasses*, designed for both animals and people and, smallest of all, *canopy crossings*, usually rope or wire structures suspended at tree canopy height to allow arboreal and flying animals such as bats, squirrels or monkeys to cross. Examples of crossing structures that are known to work include flooded culverts for toads in the Netherlands, rope bridges for sloths in Costa Rica and deep underpasses for elephants in Kenya and tigers in India.

Some green bridges are very impressive structures, big enough to provide homes for wildlife in their own right. Perhaps the biggest in the world is the Natuurbrug Zanderij Crailoo in the Netherlands, its length of 800 m allowing perhaps slightly bemused animals to cross in safety over a railway line, a business park, a river, a sports complex and, inevitably, a road.

With a few exceptions, efforts to move animals along the roadside using fences and then safely under or over the tarmac through ecoducts are extremely effective. Even such obstinate jaywalkers as frogs and toads can have their roadkill rates reduced to almost nothing by cleverly designed and judiciously placed fences and underpasses. A review of a large number of published studies on roadkill mitigation methods concluded that the combination of fencing and crossing structures reduces roadkill by an impressive 83 per cent on average, and in many cases it falls to nothing (though crossings without guiding fences perform much less well). This is a higher success rate than any other roadkill mitigation method. Animal detection systems, which automatically detect animal movements and alert drivers to their presence, have an average roadkill reduction rate of 57 per cent, and reflectors, which scatter light from car headlamps into surrounding areas to scare animals away, boast a lowly 1 per cent average success rate (though some individual studies have reported much better outcomes).

Road ecologists are increasingly turning their efforts towards under-standing how wildlife crossings can be improved and where they could most effectively be deployed. Research has shown, for instance, that animals can be encouraged to use these crossings by providing suitable habitat around their entrances, and by siting them in areas undisturbed by people and unpolluted by artificial light. There are still a number of questions to address: are wildlife crossings more effective in helping local populations of animals where the highest numbers still remain or where roadkill has reduced populations to very low levels? Is it better to place them where a few animals cross throughout the year or where there are large but short-lived seasonal movements? Is it more effective to have a small number of very large crossings or a larger number of small ones? And, importantly, how can they be designed to get the biggest conservation bang for the taxpayer's buck? As our understanding of how animals use these structures grows, so it will becomes possible to build ever more cost-effective crossings that are tailored to the needs of the target species. It is very important that we do, because wildlife crossings will become increasingly important as global warming starts to bite, creating more and more climate-change refugees fleeing across heavily trafficated landscapes in their search for conditions within which they can survive.

Wildlife crossings clearly work, but they are often expensive and not always high on road-planners' agendas, or those of their political paymasters. The 4 million miles of roads in the USA are spanned by only around 1,000 wildlife crossings, although many more are planned or under construction. Germany and the Netherlands are similarly in the process of expanding their existing networks of close to 100 wildlife bridges (and hundreds more underpasses). They are all far ahead of the UK, where just a handful of green bridges span our quarter-million miles of road.

Their scarcity is no doubt linked to their high cost, somewhere between £1 million and £10 million each on average, but against the economic costs of construction must be balanced the financial benefits of reducing roadkill. Collision with large animals usually results in costly damage to vehicles, and not infrequently injury or even death to their occupants. Serious accidents have been caused by people slamming on their brakes or swerving to avoid animals on the road. In parts of northern Spain, up to half of all road accidents are caused by collisions with animals. Researchers in Brazil have found that nearly a fifth of all the animal–vehicle collisions reported by the highway police in São Paulo state resulted in injury or death to the driver or passengers at a cost to society of over $25 million. In the USA, each collision between a car and small deer costs an average of nearly $7,000 of damage, rising to $17,000 for large deer (elk) and a whopping $30,000 for moose – and there are thousands of such collisions each year. The annual bill runs to billions of dollars.

It may be, therefore, that even the most costly road crossing structures pay for themselves over time. Researchers in Brazil collected data on more than 10,000 incidents of roadkill, finding that a fairly high proportion of these involved large animals (including threatened tapirs and giant anteaters) that could cause substantial damage to vehicles and their occupants. By calculating these costs, and balancing them against the costs of keeping larger animals off the roads in the first place, they estimated that roadkill mitigation would pay for itself in somewhere between 16 and 40 years across the entire road network, and perhaps in as little as 9 years in roadkill hotspots. The authors of a recent analysis of wildlife crossing structures in the USA concluded that it costs society less to solve the problem of wildlife-vehicle collisions than it costs to do nothing. These are purely economic costs and take no account of the pain and suffering of the animal roadkill victims or injured drivers. Perhaps driven largely by these financial considerations, politicians are starting to recognise the benefits of wildlife crossings, and the number and size of new projects is increasing, particularly in the USA. The huge $87 million Wallis Annenberg wildlife crossing, whose construction started in 2022, will span ten busy lanes of traffic that cut through the Santa Monica mountains near Los Angeles. Like its forerunners in Florida, its main aim is to help the cougar, though many other species will doubtless also benefit. The Infrastructure Investment and Jobs Act that President Joe Biden signed into law in November 2021 earmarked $350 million for animal-friendly bridges, underpasses and roadside fences.

The American bridge engineer Ted Zoli has succinctly summarised the economic arguments in favour of wildlife crossings: 'We spend $8 billion a year running over wildlife. If we took that cost and quartered it, we could build 200 animal crossings a year, and the problem of roadkill would disappear within a generation.'

Draining the countryside

Our British road network has developed piecemeal over thousands of years. It is a historic monument, the product of a long and hugely eventful past, full of quirks and anomalies, of strange bends and loops around long-disappeared features, of meandering lanes whose original destinations have been buried by time. The dense web of lines on our satnav screens traces the movements of Iron Age tribes, Roman legions, medieval drovers and seventeenth-century farmers as well as those of harassed twenty-first-century commuters. At a quarter of a million miles in length, it is very far from being a road network that is optimised for where and how we live now. If we were to redesign it from scratch to fit today's human landscape in the most efficient way, connecting the maximum number of people using the minimum length of road (as the Romans might have done), it would look very different. An optimised network would have a lower density of roads than that which permeates our ancient landscape today; the UK's current average density of around 1.7 km of road per square kilometre of land might fall to something closer to that of Spain or Ireland (1.3 km), or perhaps Portugal or Greece (0.9 km) or maybe even that of the USA (0.7 km). The third element of traffication – pervasiveness – would fall, giving wildlife some space and calm to recover over wide areas.

It is hardly likely that an entirely redesigned road system will appear any time soon, but there may be ways that we can improve the flow of traffic through our antique web of highways and byways that will bring benefits to both wildlife and people (drivers included). The idea begins by recognising that the longer our journey, the more ways there are to reach our destination. There may be only one sensible way to drive to our local shop, but two equally good ways of getting to the nearest supermarket and several more-or-less plausible ways of visiting friends who live 50 miles away. Satnavs recognise this and offer us a choice: do we want to take the route with the shortest total distance, or the one that gets us there quickest? If our chosen route is closed by roadworks or an accident, there will usually be several ways to skirt around the obstruction that add only a few minutes to our journey. What this shows is that we have redundancy in our road network, particularly away from major highways. We simply don't need all of the roads we have – so perhaps we can de-trafficate parts of our landscape, even if that means increasing traffic flow in others.

Researchers in the Netherlands have developed a concept they call the 'traffic calming rural area' (TCRA). This starts from the premise that heavily trafficated countries have dense networks of rural roads, which account for much of the fragmentation of natural habitats and most of the roadkill (and, indeed, serious accidents to people), but carry a relatively small proportion of overall vehicle volume. The TCRA, which began life as a concept to improve road safety, proposes that vehicles that use rural roads for through-transit, as opposed to local access, could instead be encouraged onto a smaller number of main roads. This could be done by reducing the capacity of some of the more redundant rural roads, for example by lowering speed limits or closing them to cars altogether, while simultaneously increasing the attractiveness to motorists of a smaller number of main through-roads, by widening them or raising the speed limit. The concept is a little like that of drainage, whereby water is removed from flooded land by improving the flow rate in surrounding rivers. This would result in the de-traffication of large areas, allowing other wildlife protection measures, such as green bridges or wildlife underpasses, to be concentrated along the busier through-roads.

The TCRA remains just a theory, backed by some computer simulations, but it is an interesting example of the kind of landscape-scale thinking that could be brought to bear if we ever decide to get to grips with the problems of road traffic in a serious way. If we want to regain some of our lost wilderness, to restore extensive areas to their previous roadless state without jamming our transport system, then concepts such as the TCRA might point the way.

De-traffication begins at home

Driving might be the most environmentally damaging thing you do. You were probably aware before you picked up this book that the convenience your car brings you comes with an environmental price tag. All those badgers and hedgehogs you see dead on the roads are daily reminders of that, and you'll have heard, too, about climate change. But you might by now be worried by just how damaging your car is to wildlife. You may not have realised that you are part of a pandemic environmental crisis that has fragmented and impoverished our wildlife like no other. Perhaps you want to do something about it. The good news is that you can. We don't need to wait for complex ideas such as TCRA to filter their way slowly through the political thought process. We can all start de-trafficating tomorrow. Here are a few suggestions:

- Keep your speed down. I will use Lewis Carroll's famous line 'What I tell you three times is true' to justify one final repetition of a core truth of road ecology: pretty much all the environmental damage you cause while driving increases exponentially with your speed.
- Avoid driving in darkness where possible, and particularly around dawn and dusk. Nocturnal animals are extremely sensitive to noise and light pollution, and are usually the most frequent victims of roadkill. Even animals that are not strictly nocturnal may be more vulnerable at night, since they often wait until after dark to cross when traffic levels are lower; large mammals and many amphibians prefer to cross the road at night and suffer their highest rates of roadkill during the hours of darkness. The periods just after dusk and just before dawn are critical, as this is when many nocturnal animals set out to travel to their feeding areas, or return from them to sleep. There are also fewer cars on the roads at night, so you will be a bigger part of the problem; the noise or light from your car might be the only thing between a bat and its meal.
- Remember that many animals travel in groups, so if you see one deer run across the road ahead of you, or one bird fly across, assume there are more following behind it and react accordingly.

- Don't drive off-road, and if driving to remote places where few other cars go, clean your car beforehand. Your car is a mobile seed bank for invasive species and you risk spreading them to new areas.
- Ensure that your tyres are correctly inflated and in good condition and that your wheels are properly aligned; avoid rapid acceleration or braking where possible. This will reduce your production of tyre wear pollution, something now known to greatly exceed in quantity and toxicity the particulate pollution that your exhaust pipe puts out. Aggressive driving can increase tyre wear pollution by a factor of a thousand.
- Go electric next time you need to buy a car (you won't have much choice in the matter soon anyway). EVs are not the answer to many of the problems of traffication but they will help a little. Consider installing solar panels on your house, if possible, to recharge it. If you do go electric, remember that your driving does not suddenly become toll-free with respect to the environment. You still need to do all the other things on this list too.
- It almost goes without saying, but here goes anyway: drive less! Think of driving as junk food and your own efforts to cut down as a diet – monitor your mileage and fuel use each month and set targets to reduce them.
- Consider joining a car-sharing scheme. Experience has shown that not only does car-sharing lower the number of cars on the roads, but it also increases vehicle occupancy, reduces per capita car travel and encourages users to spread their travel needs across other more sustainable forms of transport.
- Help road ecologists to collect valuable scientific data by joining a roadkill monitoring scheme.

Most of the car journeys we make are not even vaguely necessary and could be undertaken by other means, or not at all, without causing us any great hardship. The fact that car-sharing leads to lower per capita driving illustrates this neatly; it seems that even having to do something as trivial as booking our travel in advance is often enough to make us decide not to bother. And if that little thing is sufficient to dissuade us from jumping into the car, then the huge environmental damage caused by our driving should surely be deterrent enough.

There are, then, many ways to reduce the environmental harm caused by the three elements of traffication – volume, speed and pervasiveness – without having to give up our cars. We can prevent roadkill and restore ecological connectivity by building wildlife crossings or by perfecting systems that automatically detect animals near the road and alert approaching vehicles. We can drive slower, change road surfaces and build noise-deflecting barriers to reduce soundscape pollution, and we can re-create pristine roadless areas with little disruption to the motorist by reducing our road density in smart ways. We can protect our remaining roadless areas to ensure they stay that way. Through simple individual driving choices, we can all reduce air and water pollution and lessen our contribution to climate change. Doing all these things will also improve our health, our wealth and our happiness in ways both measurable and immeasurable. Here is another blatant repetition: de-traffication is a win–win scenario for people and wildlife.

The challenge is to decide how best to deploy these various solutions in the most effective way, reducing the greatest amount of environmental damage for the least loss of travel freedom. This is where the science of road ecology comes into its own, developing solutions and identifying where they are most needed by means of careful research. Already we know that some of the most vulnerable species are those that cross roads in vast numbers over very short periods of time; land crabs and amphibians swarm across roads during just a few weeks each year. This means that the measures needed to protect them could be equally ephemeral, causing motorists minimal disturbance. The gains could be huge; a study in Germany back in 1986 found that the combination of a few temporary road closures, some fencing and some volunteers armed with buckets saved the lives of over 130,000 amphibians of 14 different species in the space of a few weeks. We know too that many other species are more vulnerable at particular times of day or year; mammals are killed mostly at night, and many species of birds are more vulnerable in the summer than in winter. Knowing this again allows us to target actions to the most crucial times and places, leaving roads free of restrictions at other times. There are technological developments that could help too, for instance by integrating roadkill warnings into smartphones and satnav systems.

The opportunities for improving our countryside are limitless.

A glimpse of the future

There is no need for us to continue sacrificing our wildlife, our environment and our health on the altar of traffication. Neither is there a need to give up all the benefits we gain from liberated, convenient and (sometimes) economically beneficial road travel. With a little effort and willpower, and the judicious appliance of science, we can have much of the best of both worlds. A de-trafficated world is not a world without cars; it is one in which we make more of our journeys in other ways, in which the journeys we make are travelled less intrusively, more considerately of the natural world; in which the footprint of our tyres is reduced. It is a rebalanced world in which a right to drive is secondary to a right not to be harmed by the driving of others; a world in which we all have a right not to *need* to drive. It is a world in which we are willing to accept the closure of a country lane for a couple of weeks each year to allow toads to cross, or to reduce our speed along a short section of motorway where it runs through a sensitive nature reserve. It is not a Luddite, technophobe world that shuns innovation – it is the exact opposite, a world that makes smarter use of rapidly advancing technology. It is a world that sees the burning of Jurassic compost in a combustion chamber invented by bewhiskered Victorians not as an icon of modernity or as a symbol of social prestige, but as a tarnished anachronism that embarrasses our genius for betterment. It is a cleaner, quieter, healthier place for everyone to live.

Traffication has crept up on us insidiously. It is a creeping barrage of noise and pollution that has eroded and fragmented our environment, and our connections with it, so gradually that we have forgotten how much richer and more integral to our lives nature used to be. In 2020, we were given a brief reminder of what the car has cost us. The COVID-19 pandemic was a human tragedy of immeasurable proportions but it did at least reveal to us, fleetingly, what an even slightly de-trafficated world might look like. The travel restrictions put in place to slow the spread of the disease did not return us to some state of pre-industrial, or post-apocalyptic, immobility; we were not de-trafficated back to the Stone Age. Our driving fell by little more than half in most places during even the strictest restrictions, and summed over the whole of 2020, traffic volume in Britain was only a fifth lower than it had been in the previous year; in the USA it fell by just a tenth.

Yet within days of lockdown the air cleared sufficiently for mountain ranges and stars to reappear through the smog for the first time in a human generation. Satellites orbiting the planet detected a record drop in the levels of carbon dioxide in our atmosphere. The planet quite literally stopped vibrating to the rumble of vehicles: all around the world, seismologists were able to detect faint creaks and gurgles of magma deep beneath the Earth's surface that are normally swamped by the ground-bending shudder of traffic.

People started to notice birdsong; many found that this reconnection with nature helped them to cope with the isolation of lockdown. The sound they heard was the natural song of birds, not some simplified, high-pitched version screamed out over the clamour of engines. Roadkill rates fell for the first time in generations. Animals hounded by traffic into a reclusive nocturnal existence started to reclaim the day. The internet was flooded with videos of deer, bears, wild boars, otters, pumas, goats and jackals emerging nervously by the roadside, cautiously approaching suburbs. Birds and bats swooped low near newly quieted roads where none had been the year before.

It was as if nature had been given a chance to breathe out, to relax; to revisit, however briefly, ancestral homelands long lost to the car.

This, the natural world seemed to be saying to us, *is how we could live together…*

Notes

1. The King of the Road

1. The Stoners would not have known this because road ecology did not achieve recognition as a discrete scientific discipline until 1998, 20 years after Lillian's death. Dayton Stoner was, somewhat predictably for the era, listed as the article's sole author.

2. Émile Levassor fell victim to his own invention when he was seriously injured in the 1896 Paris–Marseilles race after swerving to avoid a dog. He continued the race, but never recovered from his injuries and died the following year.

3. In February 1896, Ellis gave the Prince of Wales (later King Edward VII) his first drive in a car, and the following year he achieved national fame by scaling the steep hill of Malvern Beacon in a specially built Daimler. But in later life Ellis appears to have fallen on hard times. His grand house in Datchet was sold in 1908 and in the 1911 Census he was recorded as living alone in a hotel room in Weymouth, Dorset, his occupation somewhat cruelly listed as 'loafer'. He died in 1913 at the Grand Hotel in Plymouth.

4. The precise numbers were 1,681 steam vehicles, 1,575 electric vehicles and 936 petrol vehicles.

5. Joad's career, and indeed his life, might have ended more happily had he bought a car. In 1948, he was convicted of travelling on a train without a ticket and fined £2. This caused a front-page scandal that resulted in his dismissal from the BBC, wrecked his hopes of a peerage and may have exacerbated the illness that was to kill him.

2. Traffication

6. Detroit was the epicentre of motor manufacturing in the States. Ford, Dodge, Packard and Chrysler, among others, all made the city their production base (as in Britain, many of these firms had started life as cycle makers). So synonymous was Detroit with car-making that it became known as 'Motor Town', later abbreviated to 'Motown'.

7. These stirring words no doubt expressed Ford's beliefs but were probably written for him by someone else. Ford was more a man of deeds than of fine words.

8. Different makes (or marques) often shared the same model names – Sevens, Eights and so on. These numbers refer to horsepower tax classes. Horsepower tax was introduced in 1910 (and revised in 1921) to tax powerful vehicles more heavily than light ones. Together with protective tariffs, these taxes partly protected British manufacturers from imports such as the Model T, which were

more powerful for their price than domestically produced cars. Henry Ford's response to the import tariff was simple: he built a factory at Trafford Park in Manchester in 1911 and started making his cars there.

9. Based on figures published by the US Bureau of Transportation Statistics, and interpolating for missing data, I estimate the total distance travelled by motor vehicles on the USA's roads between 1970 and 2021 at 116 trillion miles. A light year, the distance that light travels in a year of 365.25 Earth days, is 5.9 trillion miles, so America's motorists have driven something like 20 light years since the first Moon landing. Proxima Centauri, the closest member of the star cluster known as Alpha Centauri, is 4.2 light years away. There are no standardised figures for the total number of vehicle miles travelled globally, but if we assume, on the basis of vehicle numbers, that Americans do a fifth of all the world's driving, then the world's motor vehicles cover something like 15 trillion miles, or 2.5 light years, each year. Halley's Comet travels 7.6 billion miles during its lonely 76-year orbit, so we lap it nearly 2,000 times a year, or once every 4.4 hours.

3. 'An Inconspicuous Splotch of Red'

10. The secret to these calculations is to look not at the mortality rate, but at the survival probability. A one in a thousand chance of being killed per crossing is the same as a probability of 0.999 (or 99.9 per cent) of *not* being killed. If our badger makes two crossings a day, or 730 per year, the probability that it survives one year is $0.999^{730} =$ 0.482 (or 48.2 per cent), and its probability of *not* surviving is thus one minus this, or 0.518 (51.8 per cent). The probability of it dodging traffic for two years is $0.999^{(730 \times 2)} = 0.232$ (23.2 per cent), less than one in four. Of course, this simple example does not consider its chances of dying from other causes.

11. Calculated as: $0.999^{626} \times 0.995^{104} = $ a 0.317 probability of survival, hence a 0.683 (68.3 per cent) probability of death.

12. Dayton Stoner's obituary, published in the *Journal of Mammalogy* in 1945 (the year after his death at the age of just 60), noted that 'Apparently Stoner was ruled rather strictly by two compulsions – one emanating from his personal reserve, the other from his scientific discipline'; the latter, no doubt, explained his reluctance to extrapolate. The rather ungenerous obituarist, who appears never even to have met Dayton, also criticised his scientific writings for being 'unembellished by expressions of feeling', which seems a bit harsh.

13. Most studies of roadkill during the COVID-19 lockdown reported a significant drop in collision rates, but a few found that the increased use of quieter roads by animals, coupled with higher vehicle speeds in the absence of congestion, meant that roadkill rates did not always fall, and in some places even increased slightly.

14. Roadkill rates also seem to be extremely high on another island group, the Galápagos, suggesting that island species are particularly vulnerable to traffic.

4. Living with Roadkill

15. Brabazon was not a sympathetic character. In the same commentary, he asked 'Why such concern over 7,000 roads deaths? More than 6,000 people commit suicide every year, and nobody makes a fuss about that.' In 1909, he caged a

piglet in a small basket, tied it to the wing of his aeroplane, stuck on a label reading 'I am the first pig to fly' and took off. Maybe it was funny at the time.

16. Birds in the crow family (corvids) have an unusual way of getting around on land. At slower speeds they walk, but if they need to speed up they break into a unique bouncing hop. Unlike other hopping animals, the feet of a bouncing crow do not hit the ground at exactly the same time.

5. Traffic Islands and Invasion Highways

17. This is a more contentious statement than it might appear. Uzbekistan only qualifies as being double landlocked if the Caspian is not counted as a sea. The countries bordering this body of water have long argued over whether it is a lake or a sea, because the laws governing them are different. At a meeting in 2018 they all got together and sensibly decided that it is neither.

18. The results of these studies are not *entirely* unanimous but they are broadly consistent – of the twenty or so I have read, only two found no genetic divergence between road-divided populations.

19. Dung beetles were also introduced to Australia, blamelessly in this case, to clean up the mess left by herds of another introduction, the cow. Cows in Australia are often pastured in fields of perennial ryegrass, yet another non-native species. Dogs, also not native to Australia, often eat cane toads and are killed as a result. So: one non-native species is killed by eating another non-native species that preys upon another non-native species, which relies on the digested remains of another non-native species that are excreted by another non-native species. I wonder if there are any food chains longer than this that involve only introduced species.

20. Unfortunately for the native wildlife, the high death rate of cane toads on the roads does little to reduce their toxicity - the sun-dried carcass of a long-flattened cane toad can be just as poisonous as the living animal.

21. It has been suggested that in some of these regions, such as Mongolia, building roads in currently roadless areas might actually bring environmental benefits since it would concentrate traffic and prevent widespread off-road driving, a major cause of soil erosion.

6. Thunder Road

22. The emperor in question was Claudius, who was parodied in popular culture as being permanently sleepy.

23. Though he got it the wrong way around, as he banned chariots and wagons during the first ten hours of the day and so forced them to use the roads in the evening and at night. Poor Juvenal was losing sleep over this blunder a century and a half later. Chariots carrying triumphant generals, priests and vestal virgins were, naturally, exempt from any restrictions.

24. Other natural phenomena that are measured using logarithmic scales include the power of earthquakes (the Richter scale), the acidity or alkalinity of liquids (the pH scale) and the brightness of stars (the magnitude scale).

25. If you want to work out how much louder one sound is than another, say 73 dB and 70 dB for example, here's the trick: subtract the smaller from the larger (73–70 = 3), divide the result by ten (3÷10 = 0.3) and then raise 10 to the power of that number ($10^{0.3}$ = 1.995, or about twice as loud).

26. Recent research suggests that elephants and seals might also have extended learning periods when it comes to vocal communication.

27. The Ecuadorian hillstar, a gorgeous species of hummingbird with a dazzlingly iridescent purple-blue head, has recently been found to sing at 13 kHz, by far the highest-pitched bird song yet known. It is far beyond the hearing range of most other birds but within the range of many people (and, presumably, other Ecuadorian hillstars).

28. Birds do have one slight advantage when it comes to roadside living, because unlike mammals they cannot become permanently deaf (or not, at least, through over-exposure to loud noises). This is because birds are able to regenerate the sensory cells of the inner ear if they become damaged, whereas in mammals (ourselves included, of course) the cells die, leading to permanent hearing loss.

29. Because it is so easy to keep in captivity and because it sings a relatively simple song that is easy to analyse, the zebra finch has become the standard (or 'model') species used in the laboratory study of birdsong.

30. My father recalls that when, as a young boy in the 1940s, he visited his grandmother and her friends, all of them cotton-mill workers, he found their bizarrely exaggerated facial expressions very amusing. Only many years later did he realise that what he had witnessed was the last use of 'mee-mawing', a silent, mime-like form of communication that evolved in the deafeningly noisy Lancashire weaving-sheds to allow people to 'talk' over distances of 10 yards or more while working with their hands. Each mill had its own dialect.

31. From the few studies available, it seems that frogs that are able to raise the volume and pitch of their voices in the presence of traffic noise also increase the rate at which they call, whereas those that can't change their voices either reduce their calling rate or stop calling altogether.

32. The major exception is motorbikes, whose engines are not enclosed and whose exhaust pipes and silencers are short. At 30 mph motorbikes produce twice as much noise as a car, and at 50 mph they exceed the noise of trucks and buses. Speeding motorbikes can produce in excess of a dangerously high 95 dB.

33. Even the lower level of 45 dB may be an overestimate. Recent research suggests that communication between black-capped chickadees can start to break down in noise as quiet as 40 dB, and that numbers of woodland birds start to fall in the presence of just 42 dB of traffic noise.

7. Emission Creep

34. Climate change will have, and indeed is already having, catastrophic impacts on the planet's biodiversity. But because road traffic is far from being the only contributor to this threat, and because the impacts of climate change on biodiversity have been extensively discussed in other books (some are listed in the references at the back of this book), they are not covered in detail here.

35. Nitrogen oxides comprise mostly nitric oxide (NO) and a smaller percentage of the more poisonous nitrogen dioxide (NO_2); collectively they are known as NOx.

36. These size classes are not mutually exclusive: they fit inside each other like Russian dolls. PM_{10} includes both $PM_{2.5}$ and $PM_{0.1}$, and $PM_{2.5}$ includes $PM_{0.1}$.

37. At least, that's what car manufacturers claim – measurements often tell a different story. Older diesel vehicles emit less carbon dioxide per mile than

petrol cars and more nitrogen oxides and particulates, but improvements in engine efficiency and post-combustion filtering means that in new vehicles there is now much less difference in gas or particulate emissions between the two engine types.

38. Though one study from France detected elevated levels of lead further than 300 m from the roadside.

39. Or, to give it its full name, N-(1,3-dimethylbutyl)-N′-phenyl-p-phenylenedi-amine quinone.

40. Other salts commonly spread on the roads include calcium chloride and magnesium chloride, whose impacts on freshwater life appear to be similar or even more severe. Road salt also contains anti-caking agents such as sodium ferrocyanide; although it is not as toxic to humans as its name suggests, it may have impacts on other forms of life.

41. On the other hand, there is evidence that frogs living in more saline water have lower rates of infection by the deadly chytrid fungus, which has devastated amphibian populations worldwide.

42. Well, they have not changed very much, anyway. Growth rings in fossil shells suggest that during the age of dinosaurs some 70 million years ago, days may have been around 30 minutes shorter than they are now because the Earth rotated very slightly faster around its axis.

43. Apologists for the car might argue that street lighting should not form part of the environmental case against motor vehicles, because it would be there anyway; after all, Paris had electric streetlights by 1878, a decade before Bertha Benz's historic ride ushered in the age of the car. But this is likely to be true only in urban areas, where street lighting forms a fairly minor component of light pollution; it does not seem unreasonable to add street lighting in rural areas, where it makes up a high proportion of total light pollution, to the car's charge-sheet.

8. In the Zone

44. Safer, at least, to humans and other vertebrates. Many of the products that replaced DDT and the acutely poisonous organophosphates have been found to be just as harmful, if not more so, to non-target invertebrates and plants.

45. In scientific comparisons, the *control* is the sample that is not exposed to whatever it is the researchers are interested in, and is used as a baseline.

46. Two exceptions to this are occasional catastrophic droughts in the Sahel region of Africa, which cause short-term dips in the populations of birds such as white-throats and sand martins, and outbreaks of diseases such as trichomonosis and avian flu, which cause populations of vulnerable species such as greenfinches or geese to crash suddenly.

47. For any statisticians out there, the Pearson correlation coefficient for Figure 6 is -0.924, and for Figure 7 it is -0.961.

48. They may, however, take heart from research from Wisconsin, USA, which has shown that as wolves spread into a new area, the number of collisions between cars and deer falls, because wolves tend to move along roadsides, scaring deer away from them. The financial savings of the reduced collision rate with cars massively outweigh the cost of compensating farmers for wolf attacks on livestock.

9. The Sixth Horseman

49.   Pulling all the world's most threatened species one step back from extinction and protecting all the world's most important sites for wildlife would cost something in the region of $100 billion a year. We spend more than ten times this – well over $1 trillion – on soft drinks. The irony is that without wild nature, we would all die of thirst.

50.   Cuckoos are nest parasites, laying their eggs in the nests of other species and letting them do all the work of raising the chicks. Their 'reward' for all this hard work is that as soon as the cuckoo chick hatches, it kills all the host's own offspring.

51.   Unfortunately, sample sizes for these species were too small to allow them to be included in Sophia Cooke's data-hungry analyses of British birds' responses to roads. However, her results did suggest, as have several others, that long-distance migrants (which all three species are) are particularly sensitive to road noise.

10. Winners and Losers

52.   Up to a point, at least. The dataset that Sophia analysed was the BTO's Breeding Bird Survey, collected by thousands of volunteers. The selection of survey sites is semi-random – volunteers are given a small number of randomly selected sites from which to choose the one they will survey in the coming years. Any particularly heavily trafficated sites will surely be avoided – nobody studying birds for pleasure wants a study site with a motorway running through the middle of it! So what happens to numbers of birds in the most heavily trafficated sites of all remains unclear. Maybe even the wrens and the robins avoid them.

53.   This raises the interesting question of how far a native species has to spread of its own accord along artificial corridors before we start to consider it invasive.

11. Five Reasons for Hope

54.   The exception is electric motorbikes, which are very much quieter than petrol-powered ones.

55.   Indeed, we can revisit the issue of whether we need street lighting at all; roads are lit in the belief that this improves road safety and reduces crime, but neither assumption appears to have any factual basis.

56.   The Buchanan Report has been much criticised for practically ignoring the potential of cycling to meet urban transport needs. Buchanan himself was an ardent motorist.

57.   And about as harmful. A recent study from delightfully untrafficated New Zealand concluded that 'The current transport system in NZ, like many other car-dominated transport systems, has substantial negative impacts on health, at a similar level to the effects of tobacco and obesity.'

58.   Economists have long recognised a phenomenon known as the 'oil curse', a tendency for oil wealth to be associated with autocracy, corruption and war.

59.   Some motorists remain implacably, sometimes even violently, opposed to LTN restrictions, but recent research shows that twice as many people approve of them as disapprove. Car ownership in LTN areas usually falls by a fifth or more once the barriers are put in place.

60.   Though in the very same year, the very same government committed to a five-year, £27 billion roadbuilding programme with plans to construct thousands of miles of new highways. Old habits die hard.

61.   Banning the use of lead in petrol has not entirely removed the problem – lead is a highly persistent poison. A study published in 2021 suggested that a third or more of the lead in London's airborne particles derives from the burning of leaded fuel, 20 years after it was withdrawn from sale.

List of scientific names of species mentioned in the text

In cases where abbreviated common names have been employed in the text, for example where I have used 'wren' to refer to the species of wren that occurs in the UK rather than to any of the other (many) species of wrens in the world, a more explicit common name, following general usage, is given in parentheses.

Birds

American crow *Corvus brachyrhynchos*
Andean condor *Vultur gryphus*
barn owl *Tyto alba*
black lark *Melanocorypha yeltoniensis*
blackbird (common blackbird) *Turdus merula*
black-capped chickadee *Poecile atricapillus*
black-tailed godwit *Limosa limosa*
blue tit *Cyanistes caeruleus*
bluethroat *Luscinia svecica*
Bullock's oriole *Icterus bullockii*
buzzard (common buzzard) *Buteo buteo*
chaffinch (common chaffinch) *Fringilla coelebs*
cinereous vulture *Aegypius monachus*
cliff swallow *Petrochelidon pyrrhonota*
corn bunting *Emberiza calandra*
corncrake *Crex crex*
cuckoo (common cuckoo) *Cuculus canorus*
dunnock *Prunella modularis*
Florida scrub-jay *Aphelocoma coerulescens*
forest raven *Corvus tasmanicus*
golden nightjar *Caprimulgus eximius*
great bustard *Otis tarda*
great tit *Parus major*
greater prairie chicken *Tympanuchus cupido*
grey heron *Ardea cinerea*
grey partridge *Perdix perdix*
hazel grouse *Tetrastes bonasia*
house sparrow *Passer domesticus*
horned lark (or shorelark) *Eremophila alpestris*

lapwing (northern lapwing) *Vanellus vanellus*
lesser kestrel *Falco naumanni*
Levant sparrowhawk *Accipiter brevipes*
linnet (common linnet) *Linaria cannabina*
long-legged buzzard *Buteo rufinus*
meadow pipit *Anthus pratensis*
magpie (Eurasian magpie) *Pica pica*
Nechisar nightjar *Caprimulgus solala*
nightingale (common nightingale) *Luscinia megarhynchos*
northern flicker *Colaptes auratus*
northern mockingbird *Mimus polyglottos*
oilbird *Steatornis caripensis*
oystercatcher (Eurasian oystercatcher) *Haematopus ostralegus*
pallid harrier *Circus macrourus*
pheasant (common or ring-necked pheasant) *Phasianus colchicus*
pied (or white) wagtail *Motacilla alba*
raven (common raven) *Corvus corax*
red kite *Milvus milvus*
red-footed falcon *Falco vespertinus*
red-headed woodpecker *Melanerpes erythrocephalus*
red-winged blackbird *Agelaius phoeniceus*
redshank (common redshank) *Tringa totanus*
reed warbler (common reed warbler) *Acrocephalus scirpaceus*
robin (Eurasian robin) *Erithacus rubecula*
ruff *Calidris pugnax*
siskin (Eurasian siskin) *Spinus spinus*
skylark (Eurasian skylark) *Alauda arvensis*
snowy owl *Bubo scandiacus*
sociable lapwing *Vanellus gregarius*
song thrush *Turdus philomelos*
southern cassowary *Casuarius casuarius*
starling (common starling) *Sturnus vulgaris*
steppe eagle *Aquila nipalensis*
tawny owl *Strix aluco*
tufted titmouse *Baeolophus bicolor*
turkey vulture *Cathartes aura*
turtle dove *Streptopelia turtur*
tree sparrow *Passer montanus*
tree swallow *Tachycineta bicolor*
waxwing (Europe: Bohemian waxwing) *Bombycilla garrulus*
waxwing (North America: cedar waxwing) *Bombycilla cedrorum*
wheatear (northern wheatear) *Oenanthe oenanthe*
white-crowned sparrow *Zonotrichia leucophrys*
willow warbler *Phylloscopus trochilus*
woodpigeon (common woodpigeon) *Columba palumbus*
wren (Eurasian wren) *Troglodytes troglodytes*
yellow-billed cuckoo *Coccyzus americanus*
yellowhammer *Emberiza citronella*
yellow wagtail *Motacilla flava*
zebra finch *Taeniopygia (guttata) castanotis*

Mammals

American black bear *Ursus americanus*
Amur tiger *Panthera tigris altaica*
Asiatic cheetah *Acinonyx jubatus venaticus*
barbastelle bat *Barbastella barbastellus*
Bechstein's bat *Myotis bechsteinii*
bighorn sheep *Ovis canadensis*
black-tailed jackrabbit *Lepus californicus*
bobcat *Lynx rufus*
brown hyena *Parahyaena brunnea*
brown rat *Rattus norvegicus*
California ground squirrel *Otospermophilus beecheyi*
chimpanzee *Pan troglodytes*
coyote *Canis latrans*
coypu *Myocastor coypus*
dwarf mongoose *Helogale parvula*
eastern cottontail rabbit *Sylvilagus floridanus*
eastern quoll *Dasyurus viverrinus*
elk *Cervus canadensis*
Eurasian badger *Meles meles*
Eurasian beaver *Castor fiber*
Eurasian (or European) brown bear *Ursus arctos arctos*
Eurasian lynx *Lynx lynx*
Eurasian otter *Lutra lutra*
European bison *Bison bonasus*
European hare *Lepus europaeus*
European hedgehog *Erinaceus europaeus*
European polecat *Mustela putorius*
European rabbit *Oryctolagus cuniculus*
Florida panther *Puma concolor coryi*
forest elephant *Loxodonta cyclotis*
giant anteater *Myrmecophaga tridactyla*
golden lion tamarin *Leontopithecus rosalia*
grey squirrel *Sciurus carolinensis*
grizzly bear *Ursus arctos horribilis*
hazel dormouse *Muscardinus avellanarius*
Iberian lynx *Lynx pardinus*
impala *Aepyceros melampus*
koala *Phascolarctos cinereus*
leopard *Panthera pardus*
little spotted cat *Leopardus tigrinus*
maned wolf *Chrysocyon brachyurus*
moose *Alces alces*
mountain lion (or cougar) *Puma concolor*
mule deer *Odocoileus hemionus*
muntjac *Muntiacus reevesi*
northern quoll *Dasyurus hallucatus*
pine marten *Martes martes*
prairie dogs *Cynomys* species
red deer *Cervus elaphus*

red fox *Vulpes vulpes*
red squirrel *Sciurus vulgaris*
reindeer (caribou) *Rangifer tarandus*
roe deer *Capreolus capreolus*
russet ground squirrel *Spermophilus major*
South American tapir *Tapirus terrestris*
sperm whale *Physeter macrocephalus*
Stephens' kangaroo rat *Dipodomys stephensi*
striped (or common) skunk *Mephitis mephitis*
Tasmanian devil *Sarcophilus harrisii*
thirteen-lined ground squirrel *Ictidomys tridecemlineatus*
tiger *Panthera tigris*
wild boar *Sus scrofa*
wolf *Canis lupus*
wolverine *Gulo gulo*
wombat (common wombat) *Vombatus ursinus*
wood mouse *Apodemus sylvaticus*

Amphibians

African clawed frog *Xenopus laevis*
cane toad *Rhinella marina*
common frog *Rana temporaria*
common toad *Bufo bufo*
European tree frog *Hyla arborea*

Reptiles

Bahamian pygmy boa constrictor *Tropidophis canus*
Bahamian racer snake *Cubophis vudii*
black-headed python *Aspidites melanocephalus*
black ratsnake *Pantherophis obsoletus*
blue-bellied kukri *Oligodon melaneus*
copperhead *Agkistrodon contortrix*
dice snake *Natrix tessellata*
dunes sagebrush lizard *Sceloporus arenicolus*
dugite *Pseudonaja affinis*
Galápagos lava lizard *Microlophus albemarlensis*
Indo-Chinese rat snake *Ptyas korros*
timber rattlesnake *Crotalus horridus*

Fish

chum salmon *Oncorhynchus keta*
coho salmon *Oncorhynchus kisutch*

Insects and other invertebrates

adonis blue *Polyommatus bellargus*
chalkhill blue *Polyommatus coridon*

copse snail *Arianta arbustorum*
European glow-worm *Lampyris noctiluca*
grizzled skipper *Pyrgus malvae*
honey bee *Apis mellifera*
monarch butterfly *Danaus plexippus*
red imported fire ant *Solenopsis invicta*
silver-spotted skipper *Hesperia comma*

Plants

buck's-horn plantain *Plantago coronopus*
common poppy *Papaver rhoeas*
common ragweed *Ambrosia artemisiifolia*
corn buttercup *Ranunculus arvensis*
corn marigold *Glebionis segetum*
corncockle *Agrostemma githago*
crested cow-wheat *Melampyrum cristatum*
Danish scurvygrass *Cochlearia danica*
green-flowered helleborine *Epipactis phyllanthes*
lady orchid *Orchis purpurea*
lesser sea-spurrey *Spergularia marina*
lizard orchid *Himantoglossum hircinum*
red fescue *Festuca rubra*
reflexed saltmarsh-grass *Puccinellia distans*
sea plantain *Plantago maritima*
sweet pepperbush *Clethra alnifolia*
sweet vernal grass *Anthoxanthum odoratum*

References

For reasons of space, the following pages list only the key references and those covering examples discussed explicitly in the text. A more extensive bibliography is available to download from the Pelagic website (https://pelagicpublishing.com/pages/traffication-resources).

Reviews and key summaries

Altringham, J. & Kerth, G. (2016) Bats and roads. In: *Bats in the Anthropocene: Conservation of Bats in a Changing World*, ed. C.C. Voigt & T. Kingston: 35–62. Cham: Springer International Publishing.

Ament, R., Clevenger, A.P., Yu, O. & Hardy, A. (2008) An assessment of road impacts on wildlife populations in U.S. National Parks. *Environmental Management* 42: 480.

Andrews, K.M., Gibbons, J.W. & Jochimsen, D.M. (eds.) (2008) *Ecological Effects of Roads on Amphibians and Reptiles: A Literature Review*. Salt Lake City, UT: Society for the Study of Amphibians and Reptiles.

Benitez-López, A., Alkemade, R. & Verweij, P.A. (2010) The impacts of roads and other infrastructure on mammal and bird populations: a meta-analysis. *Biological Conservation* 143: 1307–1316.

Coffin, A. (2007) From roadkill to road ecology: a review of the ecological effects of roads. *Journal of Transport Geography* 15: 396–406.

Coffin, A.W., Ouren, D.S., Bettez, N.D., Borda-de-Água, L., Daniels, A.E., Grilo, C., Jaeger, J.A.G., Navarro, L.M., Preisler, H.K. & Rauschert, E.S.J. (2021) The ecology of rural roads: effects, management & research. Issues in Ecology Report 23.

Colino-Rabanal, V. & Lizana, M. (2012) Herpetofauna and roads: a review. *Basic and Applied Herpetology* 26: 5–31.

Daigle, P. (2010) A summary of the environmental impacts of roads, management responses, and research gaps: a literature review. *BC Journal of Ecosystems and Management* 10: 65–89.

Dean, W.R.J., Seymour, C.L., Joseph, G.S. & Foord, S.H. (2019) A review of the impacts of roads on wildlife in semi-arid regions. *Diversity* 11: 10.3390/d11050081.

Erritzøe, J. (2002) Bird traffic casualties and road quality for breeding birds: a summary of existing papers with a bibliography. http://www.birdresearch.dk/unilang/traffic/trafik.htm.

Fahrig, L. & Rytwinski, T. (2009) Effects of roads on animal abundance: an empirical review and synthesis. *Ecology and Society* 14: 21.

Forman, R.T.T. & Alexander, L.E. (1998) Roads and their major ecological effects. *Annual Review of Ecology and Systematics* 29: 207–231.

Harrison, R.M. & Hester, R.E. (eds.) (2017) *Environmental Impacts of Road Vehicles: Past, Present and Future (Issues in Environmental Science and Technology Volume 44)*. Royal Society of Chemistry, UK.

Johnson, C., Jones, D., Matthews, T. & Burke, M. (2022) Advancing avian road ecology research through systematic review. *Transportation Research Part D: Transport and Environment* 109: 103375.

Jones, J.A., Swanson, F.J., Wemple, B.C. & Snyder, K.U. (2000) Effects of roads on hydrology, geomorphology, and disturbance patches in stream networks. *Conservation Biology* 14: 76–85.

Kociolek, A.V., Clevenger, A.P., Clair, C.C.S. & Proppe, D.S. (2011) Effects of road networks on bird populations. *Conservation Biology* 25: 241–249.

Konstantopoulos, K., Moustakas, A. & Vogiatzakis, I.N. (2020) A spatially explicit impact assessment of road characteristics, road-induced fragmentation and noise on bird species in Cyprus. *Biodiversity* 21: 61–71.

Mammides, C., Kounnamas, C., Goodale, E. & Kadis, C. (2016) Do unpaved, low-traffic roads affect bird communities? *Acta Oecologica* 71: 14–21.

Martin, A.E., Graham, S.L., Henry, M., Pervin, E. & Fahrig, L. (2018) Flying insect abundance declines with increasing road traffic. *Insect Conservation and Diversity* 11: 608–613.

Melis, C., Olsen, C.B., Hyllvang, M., Gobbi, M., Stokke, B.G. & Røskaft, E. (2010) The effect of traffic intensity on ground beetle (Coleoptera: Carabidae) assemblages in central Sweden. *Journal of Insect Conservation* 14: 159–168.

Muñoz, P.T., Torres, F.P. & Megías, A.G. (2015) Effects of roads on insects: a review. *Biodiversity and Conservation* 24: 659–682.

Oxley, D.J., Fenton, M.B. & Carmody, G.R. (1974) The effects of roads on populations of small mammals. *Journal of Applied Ecology* 11: 51–59.

Perumal, L., New, M.G., Jonas, M. & Liu, W. (2021) The impact of roads on sub-Saharan African ecosystems: a systematic review. *Environmental Research Letters* 16: 113001.

Rytwinski, T. & Fahrig, L. (2015) The impacts of roads and traffic on terrestrial animal populations. In: *Handbook of Road Ecology*, ed. R. van der Ree, D.J. Smith & C. Grilo: 237–246. London: John Wiley & Sons.

Spellerberg, I.F. (1998) Ecological effects of roads and traffic: a literature review. *Global Ecology and Biogeography* 7: 317–333.

Spellerberg, I.F. & Morrison, T. (1998) *The Ecological Effects of New Roads: A Literature Review*. Wellington, New Zealand: Department of Conservation.

Taylor, B.D. & Goldingay, R.L. (2010) Roads and wildlife: impacts, mitigation and implications for wildlife management in Australia. *Wildlife Research* 37: 320–331.

Trombulak, S.C. & Frissell, C.A. (2000) Review of ecological effects of roads on terrestrial and aquatic communities. *Conservation Biology* 14: 18–30.

Underhill, J. & Angold, P. (2011) Effects of roads on wildlife in an intensively modified landscape. *Environmental Reviews* 8: 21–39.

van der Ree, R., Smith, D.J. & Grilo, C. (eds.) (2015) *Handbook of Road Ecology*. London: John Wiley & Sons.

Chapters 1 and 2

Appleyard, B. (2022) *The Car: The Rise and Fall of the Machine that Made the Modern World*. London: Weidenfeld & Nicolson.

Department for Transport (2018) *Road Traffic Forecasts 2018*. London: Department for Transport.

Jeremiah, D. (2007) *Representations of British Motoring*. Manchester: Manchester University Press.

Ladd, B. (2008) *Autophobia: Love and Hate in the Automotive Age*. Chicago: The University of Chicago Press.

Moran, J. (2010) *On Roads: A Hidden History*. London: Profile Books.

Morrison, K.A. & Minnis, J. (2012) *Carscapes: The Motor Car, Architecture, and Landscape in England*. New Haven, CT and London: Yale University Press.

Norton, P.D. (2011) *Fighting Traffic: The Dawn of the Motor Age in the American City*. Cambridge, MA: MIT Press.

Reid, C. (2015) *Roads Were Not Built For Cars: How Cyclists Were the First to Push for Good Roads & Became the Pioneers of Motoring*. Washington DC: Island Press.

Sloman, L., Hopkinson, L. & Taylor, I. (2017) *The Impact of Road Projects in England. Report for CPRE*. London: Transport for Quality of Life.

Standage, T. (2021) *A Brief History of Motion: From the Wheel to the Car to What Comes Next*. London: Bloomsbury.

Watts, R., Compton, R., McCammon, J., Rich, C., Wright, S., Owens, T. & Ouren, D. (2007) Roadless space of the conterminous United States. *Science* 316: 736–738.

Wells, C.W. (2014) *Car Country: An Environmental History*. Seattle: University of Washington Press.

Chapter 3: 'An Inconspicuous Splotch of Red'

Roadkill

Ascensão, F., Barrientos, R. & D'Amico, M. (2021) Wildlife collisions put a dent in road safety. *Science* 374: 1208.

Ascensão, F. & Desbiez, A.L.J. (2022) Assessing the impact of roadkill on the persistence of wildlife populations: a case study on the giant anteater. *Perspectives in Ecology and Conservation*. doi: 10.1016/j.pecon.2022.05.001.

Costa, I.M.de C., Ferreira, M.S., Mourão, C.L.B. & Bueno, C. (2022) Spatial patterns of carnivore roadkill in a high-traffic-volume highway in the endangered Brazilian Atlantic Forest. *Mammalian Biology* 102: 477–487.

da Rosa, C.A. & Bager, A. (2012) Seasonality and habitat types affect roadkill of neotropical birds. *Journal of Environmental Management* 97: 1–5.

García-Carrasco, J.-M., Tapia, W. & Muñoz, A.-R. (2020) Roadkill of birds in Galapagos Islands: a growing need for solutions. *Avian Conservation and Ecology* 15: 19.

Glista, D.J., DeVault, T.L. & DeWoody, J.A. (2007) Vertebrate road mortality predominantly impacts amphibians. *Herpetological Conservation and Biology* 3: 77–87.

González-Gallina, A., Benítez-Badillo, G., Rojas-Soto, O.R. & Hidalgo-Mihart, M.G. (2013) The small, the forgotten and the dead: highway impact on vertebrates and its implications for mitigation strategies. *Biodiversity and Conservation* 22: 325–342.

González-Suárez, M., Zanchetta Ferreira, F. & Grilo, C. (2018) Spatial and species-level predictions of road mortality risk using trait data. *Global Ecology and Biogeography* 27: 1093–1105.

Grilo, C., Ferreira, F.Z. & Revilla, E. (2015) No evidence of a threshold in traffic volume affecting road-kill mortality at a large spatio-temporal scale. *Environmental Impact Assessment Review* 55: 54–58.

Hels, T. & Buchwald, E. (2001) The effect of road kills on amphibian populations. *Biological Conservation* 99: 331–340.

Hodson, N.L. & Snow, D.W. (1965) The Road Deaths Enquiry, 1960–61. *Bird Study* 12: 90–99.

Horváth, G. & Varjú, D. (1998) Why do mayflies lay eggs on dry asphalt roads? Water-imitating horizontally polarized light reflected from asphalt attracts Ephemeroptera. *Journal of Experimental Biology* 201: 2273-2286.

Husby, M. (2016) Factors affecting road mortality in birds. *Ornis Fennica* 93: 212-224.

Jackson, H.D. & Slotow, R. (2002) A review of Afrotropical nightjar mortality, mainly road kills. *Ostrich* 73: 147-161.

Keilsohn, W., Narango, D. & Tallamy, D. (2018) Roadside habitat impacts insect traffic mortality. *Journal of Insect Conservation* 22: 10.1007/s10841-018-0051-2.

Langley, W.M., Lipps, H.W. & Theis, J.F. (1989) Responses of Kansas motorists to snake models on a rural highway. *Transactions of the Kansas Academy of Science (1903-)* 92: 43-48.

Litvaitis, J.A. & Tash, J.P. (2008) An approach toward understanding wildlife-vehicle collisions. *Environmental Management* 42: 688-697.

McCown, J., Kubilis, P., Eason, T. & Scheick, B. (2009) Effect of traffic volume on American black bears in central Florida, USA. *Ursus* 20: 39-46.

Medrano-Vizcaino, P., Grilo, C., Pinto, F.A.S., Carvalho, W.D., Melinski, R.D., Schultz, E.D. & Gonzalez-Suarez, M. (2022) Roadkill patterns in Latin American birds and mammals. *Global Ecology and Biogeography:* https://doi.org/10.1111/geb.13557.

Møller, A.P., Erritzøe, H. & Erritzøe, J. (2011) A behavioral ecology approach to traffic accidents: interspecific variation in causes of traffic casualties among birds. *Zoological Research* 32: 115-127.

Morelli, F., Benedetti, Y. & Delgado, J.D. (2020) A forecasting map of avian roadkill-risk in Europe: a tool to identify potential hotspots. *Biological Conservation* 249: 108729.

Pinto, F., Bager, A., Clevenger, A. & Grilo, C. (2018) Giant anteater (*Myrmecophaga tridactyla*) conservation in Brazil: analysing the relative effects of fragmentation and mortality due to roads. *Biological Conservation* 228: 148-157.

Raymond, S., Schwartz, A.L.W., Thomas, R.J., Chadwick, E. & Perkins, S.E. (2021) Temporal patterns of wildlife roadkill in the UK. *PLOS ONE* 16: e0258083.

Raynor, J.L., Grainger, C.A. & Parker, D.P. (2021) Wolves make roadways safer, generating large economic returns to predator conservation. *Proceedings of the National Academy of Sciences* 118: e2023251118.

Sadleir, R.M.F.S. & Linklater, W.L. (2016) Annual and seasonal patterns in wildlife road-kill and their relationship with traffic density. *New Zealand Journal of Zoology* 43: 275-291.

Santos, S.M., Mira, A., Salgueiro, P.A., Costa, P., Medinas, D. & Beja, P. (2016) Avian trait-mediated vulnerability to road traffic collisions. *Biological Conservation* 200: 122-130.

Shilling, F.M. & Waetjen, D.P. (2015) Wildlife-vehicle collision hotspots at US highway extents: scale and data source effects. *Nature Conservation* 11: 41-60.

Stoner, D. (1925) The toll of the automobile. *Science* 61: 56-57.

Stoner, D. (1936) Wildlife casualties on the highways. *Wilson Bulletin* December, 1936: 276-283.

van Langevelde, F., van Dooremalen, C. & Jaarsma, C.F. (2009) Traffic mortality and the role of minor roads. *Journal of Environmental Management* 90: 660-667.

Wagner, R.B., Brune, C.R. & Popescu, V.D. (2021) Snakes on a lane: road type and edge habitat predict hotspots of snake road mortality. *Journal for Nature Conservation* 61: 125978.

Wang, Y., Yang, Y., Han, Y., Shi, G., Zhang, L., Wang, Z., Cao, G., Zhou, H., Kong, Y., Piao, Z. & Merrow, J. (2022) Temporal patterns and factors influencing vertebrate roadkill in China. *Transportation Research Interdisciplinary Perspectives* 15: 100662.

Yamada, Y., Sasaki, H. & Harauchi, Y. (2010) Composition of road-killed insects on coastal roads around Lake Shikotsu in Hokkaido, Japan. *Journal of Rakuno Gakuen University* 34: 177–184.

Numbers of animals killed each year

Abra, F.D., Huijser, M.P., Magioli, M., Bovo, A.A.A. & Ferraz, K.M.P.M.de B. (2021) An estimate of wild mammal roadkill in São Paulo state, Brazil. *Heliyon* 7: e06015.

Arevalo, J., Honda, W., Arce-Arias, A. & Haeger, A. (2017) Spatiotemporal variation of roadkills show mass mortality events for amphibians in a highly trafficked road adjacent to a national park, Costa Rica. *Revista de biologia tropical* 65: 1261–1276.

Ashley, P.E., Kosloski, A. & Petrie, S.A. (2007) Incidence of intentional vehicle–reptile collisions. *Human Dimensions of Wildlife* 12: 137–143.

Barrientos, R., Martins, R.C., Ascensão, F., D'Amico, M., Moreira, F. & Borda-de-Água, L. (2018) A review of searcher efficiency and carcass persistence in infrastructure-driven mortality assessment studies. *Biological Conservation* 222: 146–153.

Baxter-Gilbert, J.H., Riley, J.L., Neufeld, C.J.H., Litzgus, J.D. & Lesbarrères, D. (2015) Road mortality potentially responsible for billions of pollinating insect deaths annually. *Journal of Insect Conservation* 19: 1029–1035.

Beckmann, C. & Shine, R. (2012) Do drivers intentionally target wildlife on roads? *Austral Ecology* 37: 629–632.

Bishop, C.A. & Brogan, J.M. (2013) Estimates of avian mortality attributed to vehicle collisions in Canada (Estimation de la mortalité aviaire attribuable aux collisions automobiles au Canada). *Avian Conservation and Ecology* 8(2): 2. http://dx.doi.org/10.5751/ACE-00604-080202.

Coelho, I.P., Kindel, A. & Coelho, A.V.P. (2008) Roadkills of vertebrate species on two highways through the Atlantic Forest Biosphere Reserve, southern Brazil. *European Journal of Wildlife Research* 54: 689–699.

Collinson, W.J., Parker, D.M., Bernard, R.T.F., Reilly, B.K. & Davies-Mostert, H.T. (2014) Wildlife road traffic accidents: a standardized protocol for counting flattened fauna. *Ecology and Evolution* 4: 3060–3071.

Erritzøe, J., Mazgajski, T.D. & Rejt, L. (2003) Bird casualties on European roads – a review. *Acta Ornithologica* 38: 77–93.

Groot Bruinderink, G.W.T.A. & Hazebroek, E. (1996) Ungulate traffic collisions in Europe. *Conservation Biology* 10: 1059–1067.

Guinard, E., Julliard, R. & Barbraud, C. (2012) Motorways and bird traffic casualties: carcass surveys and scavenging bias. *Biological Conservation* 147: 40–51.

Lee, T.S., Rondeau, K., Schaufele, R., Clevenger, A.P. & Duke, D. (2021) Developing a correction factor to apply to animal–vehicle collision data for improved road mitigation measures. *Wildlife Research*. https://doi.org/10.1071/WR20090.

Loss, S.R., Will, T. & Marra, P.P. (2014) Estimation of bird–vehicle collision mortality on U.S. roads. *The Journal of Wildlife Management* 78: 763–771.

Santos, R.A.L., Santos, S.M., Santos-Reis, M., Picanço de Figueiredo, A., Bager, A., Aguiar, L.M.S. & Ascensão, F. (2016) Carcass persistence and detectability: reducing the uncertainty surrounding wildlife-vehicle collision surveys. *PLOS ONE* 11: e0165608.

Secco, H., Ratton, P., Castro, E., da Lucas, P.S. & Bager, A. (2014) Intentional snake road-kill: a case study using fake snakes on a Brazilian road. *Tropical Conservation Science* 7: 561–571.

Seiler, A., Helldin, J.O. & Seiler, C. (2004) Road mortality in Swedish mammals: results of a drivers' questionnaire. *Wildlife Biology* 10: 225–233.

Soluk, D., Zercher, D. & Worthington, A. (2011) Influence of roadways on patterns of mortality and flight behavior of adult dragonflies near wetland areas. *Biological Conservation* 144: 1638–1643.

Teixeira, F.Z., Coelho, A.V.P., Esperandio, I.B. & Kindel, A. (2013) Vertebrate road mortality estimates: effects of sampling methods and carcass removal. *Biological Conservation* 157: 317–323.

Wembridge, D., Newman, M., Bright, P. & Morris, P. (2016) An estimate of the annual number of hedgehog (*Erinaceus europaeus*) road casualties in Great Britain. *Mammal Communications* 2: 8–14.

Impacts of roadkill

Aresco, M.J. (2005) The effect of sex-specific terrestrial movements and roads on the sex ratio of freshwater turtles. *Biological Conservation* 123: 37–44.

Barrientos, R., Ascensão, F., D'Amico, M., Grilo, C. & Pereira, H. (2021) The lost road: do transportation networks imperil wildlife population persistence? *Perspectives in Ecology and Conservation* 19: 411–416.

Coulson, G. (1997) Male bias in road-kills of macropods. *Wildlife Research* 24: 21–25.

Erickson, W., Johnson, G. & Young, D. (2005) *A Summary and Comparison of Bird Mortality from Anthropogenic Causes with an Emphasis on Collisions*. USDA Forest Service General Technical Report PSW-GTR-191.

Gibbs, J. & Shriver, G. (2002) Estimating the effects of road mortality on turtle populations. *Conservation Biology* 16: 1647–1652.

Grilo, C., Koroleva, E., Andrášik, R., Bíl, M. & González-Suárez, M. (2020) Roadkill risk and population vulnerability in European birds and mammals. *Frontiers in Ecology and the Environment* 18: 323–328.

Grilo, C., Borda-de-Água, L., Beja, P., Goolsby, E., Soanes, K., le Roux, A., Koroleva, E., Ferreira, F.Z., Gagné, S.A., Wang, Y. & González-Suárez, M. (2021) Conservation threats from roadkill in the global road network. *Global Ecology and Biogeography* 30: 2200–2210.

Huijser, M.P. & Bergers, P.J.M. (2000) The effect of roads and traffic on hedgehog (*Erinaceus europaeus*) populations. *Biological Conservation* 95: 111–116.

Jones, M.E. (2000) Road upgrade, road mortality and remedial measures: impacts on a population of eastern quolls and Tasmanian devils. *Wildlife Research* 27: 289–296.

Kerley, L.L., Goodrich, J.M., Miquelle, D.G., Smirnov, E.N., Quigley, H.B. & Hornocker, M.G. (2002) Effects of roads and human disturbance on Amur tigers. *Conservation Biology* 16: 97–108.

Lehtonen, T.K., Babic, N.L., Piepponen, T., Valkeeniemi, O., Borshagovski, A.-M. & Kaitala, A. (2021) High road mortality during female-biased larval dispersal in an iconic beetle. *Behavioral Ecology and Sociobiology* 75: 26.

Ryu, M. & Kim, J.G. (2020) Influence of roadkill during breeding migration on the sex ratio of land crab (*Sesarma haematoche*). *Journal of Ecology and Environment* 44: 23.

Chapter 4: Living with Roadkill

Evolution and evasion on the road

Brady, S.P. & Richardson, J.L. (2017) Road ecology: shifting gears toward evolutionary perspectives. *Frontiers in Ecology and the Environment* 15: 91–98.

Brieger, F., Kämmerle, J.-L., Hagen, R. & Suchant, R. (2022) Behavioural reactions to oncoming vehicles as a crucial aspect of wildlife–vehicle collision risk in three common wildlife species. *Accident Analysis & Prevention* 168: 106564.

Brown, C. & Brown, M. (2013) Where has all the road kill gone? *Current Biology* 23: R233–4.

DeVault, T.L., Blackwell, B.F., Seamans, T.W., Lima, S.L. & Fernández-Juricic, E. (2014) Effects of vehicle speed on flight initiation by Turkey vultures: implications for bird–vehicle collisions. *PLoS One* 9: e87944.

DeVault, T.L., Blackwell, B.F., Seamans, T.W., Lima, S.L. & Fernández-Juricic, E. (2015) Speed kills: ineffective avian escape responses to oncoming vehicles. *Proceedings of the Royal Society of London (Series B: Biological Sciences)* 282: 20142188.

Legagneux, P. & Ducatez, S. (2013) European birds adjust their flight initiation distance to road speed limits. *Biology Letters* 9: 20130417.

Mukherjee, S., Ray-Mukherjee, J. & Sarabia, R. (2013) Behaviour of American crows when encountering an oncoming vehicle. *Canadian Field Naturalist* 127: 229–233.

Mumme, R.L., Schoech, S.J., Woolfenden, G.W. & Fitzpatrick, J.W. (2000) Life and death in the fast lane: demographic consequences of road mortality in the Florida scrub-jay. *Conservation Biology* 14: 501–512.

Owen, D.A.S., Carter, E.T., Holding, M.L., Islam, K. & Moore, I.T. (2014) Roads are associated with a blunted stress response in a North American pit viper. *General and Comparative Endocrinology* 202: 87–92.

Roads as sources of data

Adams, C.E. (1983) Road-killed animals as resources for ecological studies. *The American Biology Teacher* 45: 256–261.

Brockie, R.E., Sadleir, R.M.F.S. & Linklater, W.L. (2009) Long-term wildlife road-kill counts in New Zealand. *New Zealand Journal of Zoology* 36: 123–134.

Canova, L. & Balestrieri, A. (2019) Long-term monitoring by roadkill counts of mammal populations living in intensively cultivated landscapes. *Biodiversity and Conservation* 28: 97–113.

Case, R.M. (1978) Interstate highway road-killed animals: a data source for biologists. *Wildlife Society Bulletin* 6: 8–13.

Cleuren, S.G.C., Patterson, M.B., Hocking, D.P., Warburton, N.M. & Evans, A.R. (2022) Fang shape varies with ontogeny and sex in the venomous elapid snake *Pseudonaja affinis*. *Journal of Morphology* 283: 287–295.

Das, A., Gower, D., Narayanan, S., Pal, S., Boruah, B., Magar, S., Das, S., Moulick, S. & Deepak, V. (2022) Rediscovery and systematics of the rarely encountered blue-bellied kukri snake (*Oligodon melaneus* Wall, 1909) from Assam, India. *Zootaxa* 5138: 417–430.

George, L., Macpherson, J., Balmforth, Z. & Bright, P. (2011) Using the dead to monitor the living: can road kill counts detect trends in mammal abundance? *Applied Ecology and Environmental Research* 9: 27–42.

Ham, C.-H., Park, S.-M., Lee, J.-E., Park, J., Lee, D.-H. & Sung, H.-C. (2022) First report of the black-headed python (*Aspidites melanocephalus* Krefft, 1864) found in the wild in the Republic of Korea. *BioInvasions Records* 11.

Kolenda, K., Kaczmarski, M., Najbar, A., Rozenblut-Kościsty, B., Chmielewska, M. & Najbar, B. (2018) Road-killed toads as a non-invasive source to study age structure of spring migrating population. *European Journal of Wildlife Research* 65: 5.

Medrano-Vizcaíno, P. & Espinosa, S. (2021) Geography of roadkills within the Tropical Andes Biodiversity Hotspot: poorly known vertebrates are part of the toll. *Biotropica* 53: 820–830.

Meek, R. (2021) Population trends of four species of amphibians in western France; results from a 15 year time series derived from road mortality counts. *Acta Oecologica* 110: 103713.

Schwartz, A.L.W., Shilling, F.M. & Perkins, S.E. (2020) The value of monitoring wildlife roadkill. *European Journal of Wildlife Research* 66: 18.

Citizen science schemes
Bíl, M., Heigl, F., Janoška, Z., Vercayie, D. & Perkins, S.E. (2020) Benefits and challenges of collaborating with volunteers: examples from national wildlife roadkill reporting systems in Europe. *Journal for Nature Conservation* 54: 125798.

Chyn, K., Lin, T.-E., Chen, Y.-K., Chen, C.-Y. & Fitzgerald, L.A. (2019) The magnitude of roadkill in Taiwan: patterns and consequences revealed by citizen science. *Biological Conservation* 237: 317–326.

Cosentino, B.J., Marsh, D.M., Jones, K.S., Apodaca, J.J., Bates, C., Beach, J., Beard, K.H., Becklin, K., Bell, J.M., Crockett, C., Fawson, G., Fjelsted, J., Forys, E.A., Genet, K.S., Grover, M., Holmes, J., Indeck, K., Karraker, N.E., Kilpatrick, E.S., Langen, T.A., Mugel, S.G., Molina, A., Vonesh, J.R., Weaver, R.J. & Willey, A. (2014) Citizen science reveals widespread negative effects of roads on amphibian distributions. *Biological Conservation* 180: 31–38.

Englefield, B., Starling, M., Wilson, B., Roder, C. & McGreevy, P. (2020) The Australian Roadkill Reporting Project – applying integrated professional research and citizen science to monitor and mitigate roadkill in Australia. *Animals* 10: 1112.

Grilo, C., Coimbra, M., Cerqueira, R., Barbosa, P., Dornas, R., Gonçalves, L., Teixeira, F., Coelho, I., Schmidt, B., Pacheco, D., Schuck, G., Esperando, I., Anza, J., Beduschi, J., Oliveira, N., Pinheiro, P., Bager, A., Secco, H., Guerreiro, M. & Kindel, A. (2018) BRAZIL ROAD-KILL: a data set of wildlife terrestrial vertebrate road-kills. *Ecology* 99: 2625.

Mayadunnage, S., Stannard, H., West, P. & Old, J. (2022) Identification of hotspots and the factors affecting wombat vehicle collisions using the citizen science tool, WomSAT. *Australian Mammalogy*.

Olson, D.D., Bissonette, J.A., Cramer, P.C., Green, A.D., Davis, S.T., Jackson, P.J. & Coster, D.C. (2014) Monitoring wildlife-vehicle collisions in the information age: how smartphones can improve data collection. *PLOS ONE* 9: e98613.

Shilling, F., Collinson, W., Bil, M., Vercayie, D., Heigl, F., Perkins, S.E. & MacDougall, S. (2020) Designing wildlife–vehicle conflict observation systems to inform ecology and transportation studies. *Biological Conservation* 251: 108797.

Sterrett, S.C., Katz, R.A., Fields, W.R. & Campbell Grant, E.H. (2019) The contribution of road-based citizen science to the conservation of pond-breeding amphibians. *Journal of Applied Ecology* 56: 988–995.

Swinnen, K., Jacobs, A., Claus, K., Ruyts, S., Vercayie, D., Lambrechts, J. & Herremans, M. (2022) 'Animals under wheels': wildlife roadkill data collection by citizen scientists as a part of their nature recording activities. *Nature Conservation* 47: 121–153.

Tiedeman, K., Hijmans, R.J., Mandel, A., Waetjen, D.P. & Shilling, F. (2019) The quality and contribution of volunteer collected animal vehicle collision data in ecological research. *Ecological Indicators* 106: 105431.

Unger, S. (2022) On the road again: touring iNaturalist for roadkill observations as a new tool for ecologists. *Journal of Wildlife and Biodiversity* 6: 72–86.

Valerio, F., Basile, M. & Balestrieri, R. (2021) The identification of wildlife–vehicle collision hotspots: citizen science reveals spatial and temporal patterns. *Ecological Processes* 10: 6.

Vercayie, D. & Herremans, M. (2015) Citizen science and smartphones take roadkill monitoring to the next level. *Nature Conservation* 11: 29–40.

Waetjen, D.P. & Shilling, F.M. (2017) Large extent volunteer roadkill and wildlife observation systems as sources of reliable data. *Frontiers in Ecology and Evolution,* 5, 10.3389/fevo.2017.00089.

Chapter 5: Traffic Islands and Invasion Highways
Road-crossing behaviour

Andrews, K.M. & Gibbons, J.W. (2005) How do highways influence snake movement? Behavioral responses to roads and vehicles. *Copeia* 2005: 772–782.

Bennett, V.J. & Zurcher, A.A. (2013) When corridors collide: road-related disturbance in commuting bats. *The Journal of Wildlife Management* 77: 93–101.

Brock, R.E. & Kelt, D.A. (2004) Influence of roads on the endangered Stephens' kangaroo rat (*Dipodomys stephensi*): are dirt and gravel roads different? *Biological Conservation* 118: 633–640.

Chen, H.L. & Koprowski, J.L. (2019) Can we use body size and road characteristics to anticipate barrier effects of roads in mammals? A meta-analysis. *Hystrix, the Italian Journal of Mammalogy* 30: 1–7.

Cibot, M., Bortolamiol, S., Seguya, A. & Krief, S. (2015) Chimpanzees facing a dangerous situation: a high-traffic asphalted road in the Sebitoli area of Kibale National Park, Uganda. *American Journal of Primatology* 77: 890–900.

Clark, B.K., Clark, B.S., Johnson, L.A. & Haynie, M.T. (2001) Influence of roads on movements of small mammals. *The Southwestern Naturalist* 46: 338–344.

Dániel-Ferreira, J., Berggren, Å., Wissman, J. & Öckinger, E. (2022) Road verges are corridors and roads barriers for the movement of flower-visiting insects. *Ecography*, e05847.

de Oliveira, P.R.R., Alberts, C.C. & Francisco, M.R. (2011) Impact of road clearings on the movements of three understory insectivorous bird species in the Brazilian Atlantic Forest. *Biotropica* 43: 628–632.

Friebe, K., Steffens, T., Schulz, B., Valqui, J., Reck, H. & Hartl, G. (2018) The significance of major roads as barriers and their roadside habitats as potential corridors for hazel dormouse migration – a population genetic study. *Folia Zoologica* 67: 98–109.

Galantinho, A., Santos, S., Eufrázio, S., Silva, C., Carvalho, F., Alpizar-Jara, R. & Mira, A. (2022) Effects of roads on small-mammal movements: opportunities and risks of vegetation management on roadsides. *Journal of Environmental Management* 316: 115272.

Goosem, M. (2002) Effects of tropical rainforest roads on small mammals: fragmentation, edge effects and traffic disturbance. *Wildlife Research* 29: 277–289.

Grilo, C., Molina-Vacas, G., Fernández Aguilar, X., Rodríguez-Ruiz, J., Ramiro, V., Porto-Peter, F., Ascensão, F., Roman, J. & Revilla, E. (2018) Species-specific movement traits and specialization determine the spatial responses of small mammals towards roads. *Landscape and Urban Planning* 169: 199–207.

Jaeger, J.A.G., Bowman, J., Brennan, J., Fahrig, L., Bert, D., Bouchard, J., Charbonneau, N., Frank, K., Gruber, B. & von Toschanowitz, K.T. (2005) Predicting when animal populations are at risk from roads: an interactive model of road avoidance behavior. *Ecological Modelling* 185: 329–348.

Laurance, S.G.W. (2004) Responses of understory rain forest birds to road edges in central Amazonia. *Ecological Applications* 14: 1344–1357.

Londe, D.W., Elmore, R.D., Davis, C.A., Hovick, T.J., Fuhlendorf, S.D. & Rutledge, J. (2022) Why did the chicken not cross the road? Anthropogenic development influences the movement of a grassland bird. *Ecological Applications* 32: e2543.

Madden, J.R. & Perkins, S.E. (2017) Why did the pheasant cross the road? Long-term road mortality patterns in relation to management changes. *Royal Society Open Science* 4: 170617–170617.

Mazerolle, M., Huot, M. & Gravel, M. (2005) Behavior of amphibians on the road in response to car traffic. *Herpetologica* 61: 380–388.

Ryu, M. & Kim, J.G. (2021) Coastal road mortality of land crab during spawning migration. *Scientific Reports* 11: 6702.

Shine, R., Lemaster, M., Wall, M., Langkilde, T. & Mason, R. (2013) Why did the snake cross the road? Effects of roads on movement and location of mates by garter snakes (*Thamnophis sirtalis parietalis*). *Ecology & Society* 9: 9.

Siers, S.R., Reed, R.N. & Savidge, J.A. (2016) To cross or not to cross: modeling wildlife road crossings as a binary response variable with contextual predictors. *Ecosphere* 7: e01292.

Thurfjell, H., Spong, G., Olsson, M. & Ericsson, G. (2015) Avoidance of high traffic levels results in lower risk of wild boar-vehicle accidents. *Landscape and Urban Planning* 133: 98–104.

Wadey, J., Beyer, H.L., Saaban, S., Othman, N., Leimgruber, P. & Campos-Arceiz, A. (2018) Why did the elephant cross the road? The complex response of wild elephants to a major road in Peninsular Malaysia. *Biological Conservation* 218: 91–98.

Wattles, D.W., Zeller, K.A. & DeStefano, S. (2018) Response of moose to a high-density road network. *The Journal of Wildlife Management* 82: 929–939.

Wilkins, K.T. (1982) Highways as barriers to rodent dispersal. *Southwestern Naturalist* 27: 459–460.

Wilson, R.R., Parrett, L.S., Joly, K. & Dau, J.R. (2016) Effects of roads on individual caribou movements during migration. *Biological Conservation* 195: 2–8.

Roads and fragmentation

Alexander, S.M., Waters, N.M. & Paquet, P.C. (2005) Traffic volume and highway permeability for a mammalian community in the Canadian Rocky Mountains. *The Canadian Geographer / Le Géographe canadien* 49: 321–331.

Andersson, P., Koffman, A., Sjödin, N.E. & Johansson, V. (2017) Roads may act as barriers to flying insects: species composition of bees and wasps differs on two sides of a large highway. *Nature Conservation* 18: 47–59.

Andrews, A. (1990) Fragmentation of habitat by roads and utility corridors: a review. *Australian Zoologist* 26: 130–141.

Arens, P., van der Sluis, T., Westende, W., Vosman, B., Vos, C. & Smulders, M.J.M. (2007) Genetic population differentiation and connectivity among fragmented moor frog (*Rana arvalis*) populations in The Netherlands. *Landscape Ecology* 22: 1489–1500.

Ascensão, F., Mata, C., Malo, J.E., Ruiz-Capillas, P., Silva, C., Silva, A.P., Santos-Reis, M. & Fernandes, C. (2016) Disentangle the causes of the road barrier effect in small mammals through genetic patterns. *PLOS ONE* 11: e0151500.

Balkenhol, N. & Waits, L.P. (2009) Molecular road ecology: exploring the potential of genetics for investigating transportation impacts on wildlife. *Molecular Ecology* 18: 4151–4164.

Baur, A. & Baur, B. (1990) Are roads barriers to dispersal in the land snail *Arianta arbustorum*? *Canadian Journal of Zoology* 68: 613–617.

Bhattacharya, M., Primack, R.B. & Gerwein, J. (2003) Are roads and railroads barriers to bumblebee movement in a temperate suburban conservation area? *Biological Conservation* 109: 37–45.

Blake, S., Deem, S.L., Strindberg, S., Maisels, F., Momont, L., Isia, I.-B., Douglas-Hamilton, I., Karesh, W.B. & Kock, M.D. (2008) Roadless wilderness area determines forest elephant movements in the Congo Basin. *PLOS ONE* 3: e3546.

Borgardt, T., Hill, K. & Crother, B. (2021) The spatial ecology of the timber rattlesnake (*Crotalus horridus*) in southeastern Louisiana. *Authorea*: 10.22541/au.163899383.33938357/v1.

Brehme, C.S., Tracey, J.A., McClenaghan, L.R. & Fisher, R.N. (2013) Permeability of roads to movement of scrubland lizards and small mammals. *Conservation Biology* 27: 710–720.

Ceia-Hasse, A., Navarro, L.M., Borda-de-Água, L. & Pereira, H.M. (2018) Population persistence in landscapes fragmented by roads: disentangling isolation, mortality, and the effect of dispersal. *Ecological Modelling* 375: 45–53.

Clark, R.W., Brown, W.S., Stechert, R. & Zamudio, K.R. (2010) Roads, interrupted dispersal, and genetic diversity in timber rattlesnakes. *Conservation Biology* 24: 1059–69.

Clarke, G.P., White, P.C.L. & Harris, S. (1998) Effects of roads on badger *Meles meles* populations in south-west England. *Biological Conservation* 86: 117–124.

Clevenger, A.P. & Kociolek, A.V. (2013) Potential impacts of highway median barriers on wildlife: state of the practice and gap analysis. *Environmental Management* 52: 1299–312.

Develey, P.F. & Stouffer, P.C. (2001) Effects of roads on movements by understory birds in mixed-species flocks in Central Amazonian Brazil. *Conservation Biology* 15: 1416–1422.

Eigenbrod, F., Hecnar, S. & Fahrig, L. (2008) Accessible habitat: an improved measure of the effects of habitat loss and roads on wildlife populations. *Landscape Ecology* 23: 159–168.

Epps, C.W., Palsbøll, P.J., Wehausen, J.D., Roderick, G.K., Ramey Ii, R.R. & McCullough, D.R. (2005) Highways block gene flow and cause a rapid decline in genetic diversity of desert bighorn sheep. *Ecology Letters* 8: 1029–1038.

Epps, C.W., Crowhurst, R.S. & Nickerson, B.S. (2018) Assessing changes in functional connectivity in a desert bighorn sheep metapopulation after two generations. *Molecular Ecology* 27: 2334–2346.

Fernandes, N., Ferreira, E.M., Pita, R., Mira, A. & Santos, S.M. (2022) The effect of habitat reduction by roads on space use and movement patterns of an endangered species, the Cabrera vole *Microtus cabrerae*. *Nature Conservation* 47: 177–196.

Fitch, G. & Vaidya, C. (2021) Roads pose a significant barrier to bee movement, mediated by road size, traffic and bee identity. *Journal of Applied Ecology* 58: 1177–1186.

Frantz, A.C., Bertouille, S., Eloy, M.C., Licoppe, A., Chaumont, F. & Flamand, M.C. (2012) Comparative landscape genetic analyses show a Belgian motorway to be a gene flow barrier for red deer (*Cervus elaphus*), but not wild boars (*Sus scrofa*). *Molecular Ecology* 21: 3445–3457.

Garcia-Gonzalez, C., Campo, D., Pola, I.G. & Garcia-Vazquez, E. (2012) Rural road networks as barriers to gene flow for amphibians: species-dependent mitigation by traffic calming. *Landscape and Urban Planning* 104: 171–180.

Gerlach, G. & Musolf, K. (2000) Fragmentation of landscape as a cause for genetic subdivision in bank voles. *Conservation Biology* 14: 1066–1074.

Goosem, M. (2001) Effects of tropical rainforest roads on small mammals: inhibition of crossing movements. *Wildlife Research* 28: 351–364.

Goosem, M. (2002) Effects of tropical rainforest roads on small mammals: fragmentation, edge effects and traffic disturbance. *Wildlife Research* 29: 277–289.

Goosem, M. (2007) Fragmentation impacts caused by roads through rainforests. *Current Science* 93.

Hennessy, C., Tsai, C.-C., Anderson, S.J., Zollner, P.A. & Rhodes Jr, O.E. (2018) What's stopping you? Variability of interstate highways as barriers for four species of terrestrial rodents. *Ecosphere* 9: e02333.

Holderegger, R. & Di Giulio, M. (2010) The genetic effects of roads: a review of empirical evidence. *Basic and Applied Ecology* 11: 522–531.

Jackson, N.D. & Fahrig, L. (2011) Relative effects of road mortality and decreased con-nectivity on population genetic diversity. *Biological Conservation* 144: 3143–3148.

Karlson, M. & Mörtberg, U. (2015) A spatial ecological assessment of fragmenta-tion and disturbance effects of the Swedish road network. *Landscape and Urban Planning* 134: 53–65.

Keller, I. & Largiadèr, C.R. (2003) Recent habitat fragmentation caused by major roads leads to reduction of gene flow and loss of genetic variability in ground beetles. *Proceedings of the Royal Society of London. Series B: Biological Sciences* 270: 417–423.

Keller, I., Nentwig, W. & Largiader, C.R. (2004) Recent habitat fragmentation due to roads can lead to significant genetic differentiation in an abundant flightless ground beetle. *Molecular Ecology* 13: 2983–2994.

Kramer-Schadt, S., Revilla, E., Wiegand, T. & Breitenmoser, U.R.S. (2004) Fragmented landscapes, road mortality and patch connectivity: modelling influences on the dispersal of Eurasian lynx. *Journal of Applied Ecology* 41: 711–723.

Kuehn, R., Hindenlang, K.E., Holzgang, O., Senn, J., Stoeckle, B. & Sperisen, C. (2007) Genetic effect of transportation infrastructure on roe deer populations (*Capreolus capreolus*). *Journal of Heredity* 98: 13–22.

Laurance, S.G.W., Stouffer, P.C. & Laurance, W.E. (2004) Effects of road clearings on movement patterns of understory rainforest birds in central Amazonia. *Conserva-tion Biology* 18: 1099–1109.

Laurance, W.F., Lovejoy, T.E., Vasconcelos, H.L., Bruna, E.M., Didham, R.K., Stouffer, P.C., Gascon, C., Bierregaard, R.O., Laurance, S.G. & Sampaio, E. (2002) Ecosystem decay of Amazonian forest fragments: a 32-year investigation. *Conservation Biology* 16: 605–618.

Lesbarrères, D., Primmer, C.R., Lodé, T. & Merilä, J. (2006) The effects of 20 years of highway presence on the genetic structure of *Rana dalmatina* populations. *Écoscience* 13: 531–538.

Lesbarrères, D. & Fahrig, L. (2012) Measures to reduce population fragmentation by roads: what has worked and how do we know? *Trends in Ecology & Evolution* 27: 374–380.

MacPherson, D., Macpherson, J. & Morris, P. (2011) Rural roads as barriers to the movements of small mammals. *Applied Ecology and Environmental Research* 9: 167–180.

Mader, H.J., Schell, C. & Kornacker, P. (1990) Linear barriers to arthropod movements in the landscape. *Biological Conservation* 54: 209–222.

Marsh, D.M., Milam, G.S., Gorham, N.P. & Beckman, N.G. (2005) Forest roads as partial barriers to terrestrial salamander movement. *Conservation Biology* 19: 2004–2008.

Marsh, D.M., Page, R.B., Hanlon, T.J., Corritone, R., Little, E.C., Seifert, D.E. & Cabe, P.R. (2008) Effects of roads on patterns of genetic differentiation in red-backed salamanders, *Plethodon cinereus*. *Conservation Genetics* 9: 603–613.

McDonald, W.R. & St. Clair, C.C. (2004) The effects of artificial and natural barriers on the movement of small mammals in Banff National Park, Canada. *Oikos* 105: 397–407.

McGregor, R.L., Bender, D.J. & Fahrig, L. (2008) Do small mammals avoid roads because of the traffic? *Journal of Applied Ecology* 45: 117–123.

Niu, H., Peng, C., Chen, Z., Wang, Z. & Zhang, H. (2021) Country roads as barriers to rodent-mediated seed dispersal in a warm-temperate forest: implications for forest fragmentation. *European Journal of Forest Research* 140: 1–12.

Noordijk, J., Prins, D., Jonge & Vermeulen, R. (2006) Impact of a road on the movements of two ground beetle species (Coleoptera: Carabidae). *Entomologica Fennica* 17: 276–283.

O'Hagan, M.J.H., McCormick, C.M., Collins, S.F., McBride, K.R. & Menzies, F.D. (2021) Are major roads effective barriers for badger (*Meles meles*) movements? *Research in Veterinary Science* 138: 49–52.

Proctor, M.F., McLellan, B.N., Strobeck, C. & Barclay, R.M.R. (2005) Genetic analysis reveals demographic fragmentation of grizzly bears yielding vulnerably small populations. *Proceedings of the Royal Society: Biological Sciences* 272: 2409–2416.

Reh, W. & Seitz, A. (1990) The influence of land use on the genetic structure of populations of the common frog *Rana temporaria*. *Biological Conservation* 54: 239–249.

Richardson, J., Shore, R., Treweek, J. & Larkin, S. (2009) Are major roads a barrier to small mammals? *Journal of Zoology* 243: 840–846.

Rico, A., Kindlmann, P. & Sedláček, F. (2009) Can the barrier effect of highways cause genetic subdivision in small mammals? *Acta Theriologica* 54: 297–310.

Riley, S.P.D., Pollinger, J.P., Sauvajot, R.M., York, E.C., Bromley, C., Fuller, T.K. & Wayne, R.K. (2006) A southern California freeway is a physical and social barrier to gene flow in carnivores. *Molecular Ecology* 15: 1733–1741.

Rondinini, C. & Doncaster, C.P. (2002) Roads as barriers to movement for hedgehogs. *Functional Ecology* 16: 504–509.

Scrafford, M.A., Avgar, T., Heeres, R. & Boyce, M.S. (2018) Roads elicit negative movement and habitat-selection responses by wolverines (*Gulo gulo luscus*). *Behavioral Ecology* 29: 534–542.

Shepard, D.B., Kuhns, A.R., Dreslik, M.J. & Phillips, C.A. (2008) Roads as barriers to animal movement in fragmented landscapes. *Animal Conservation* 11: 288–296.

Thatte, P., Joshi, A., Vaidyanathan, S., Landguth, E. & Ramakrishnan, U. (2018) Maintaining tiger connectivity and minimizing extinction into the next century: insights from landscape genetics and spatially-explicit simulations. *Biological Conservation* 218: 181–191.

Waller, J.S. & Servheen, C. (2005) Effects of transportation infrastructure on grizzly bears in northwestern Montana. *The Journal of Wildlife Management* 69: 985–1000.

Invasions along roads

Ansong, M. & Pickering, C. (2013) Are weeds hitchhiking a ride on your car? A systematic review of seed dispersal on cars. *PLOS ONE* 8: e80275.

Brown, G., Phillips, B., Webb, J. & Shine, R. (2006) Toad on the road: use of roads as dispersal corridors by cane toads (*Bufo marinus*) at an invasion front in tropical Australia. *Biological Conservation* 133: 88–94.

Brunker, K., Lemey, P., Marston, D.A., Fooks, A.R., Lugelo, A., Ngeleja, C., Hampson, K. & Biek, R. (2018) Landscape attributes governing local transmission of an endemic zoonosis: rabies virus in domestic dogs. *Molecular Ecology* 27: 773–788.

Cameron, E.K. & Bayne, E.M. (2009) Road age and its importance in earthworm invasion of northern boreal forests. *Journal of Applied Ecology* 46: 28–36.

Christen, D.C. & Matlack, G.R. (2009) The habitat and conduit functions of roads in the spread of three invasive plant species. *Biological Invasions* 11: 453–465.

Crossland, M., Brown, G. & Shine, R. (2011) The enduring toxicity of road-killed cane toads (*Rhinella marina*). *Biological Invasions* 13: 2135–2145.

Dar, P., Reshi, Z. & Shah, M. (2013) Roads act as corridors for the spread of alien plant species in the mountainous regions: a case study of Kashmir Valley, India. *Tropical Ecology* 56: 49–56.

David, A.M., Emily, S.J.R., Andrea, N.N. & Brian, P.J. (2009) Forest roads facilitate the spread of invasive plants. *Invasive Plant Science and Management* 2: 191–199.

Fernandez Murillo, M.d.P., Rico, A. & Kindlmann, P. (2015) Exotic plants along roads near La Paz, Bolivia. *Weed Research* 55: 565–573.

Follak, S., Eberius, M., Essl, F., Fürdős, A., Sedlacek, N. & Trognitz, F. (2018) Invasive alien plants along roadsides in Europe. *EPPO Bulletin* 48: 256–265.

Forys, E.A., Allen, C.R. & Wojcik, D.P. (2002) Influence of the proximity and amount of human development and roads on the occurrence of the red imported fire ant in the lower Florida Keys. *Biological Conservation* 108: 27–33.

Gelbard, J.L. & Belnap, J. (2003) Roads as conduits for exotic plant invasions in a semiarid landscape. *Conservation Biology* 17: 420–432.

Joly, M., Bertrand, P., Gbangou, R.Y., White, M.C., Dubé, J. & Lavoie, C. (2011) Paving the way for invasive species: road type and the spread of common ragweed (*Ambrosia artemisiifolia*). *Environmental Management* 48: 514–522.

Lanner, J., Dubos, N., Geslin, B., Leroy, B., Hernández-Castellano, C., Dubaić, J.B., Bortolotti, L., Calafat, J.D., Ćetković, A., Flaminio, S., Le Féon, V., Margalef-Marrase, J., Orr, M., Pachinger, B., Ruzzier, E., Smagghe, G., Tuerlings, T., Vereecken, N. & Meimberg, H. (2022) On the road: anthropogenic factors drive the invasion risk of a wild solitary bee species. *Science of The Total Environment* 827: 154246.

Lázaro-Lobo, A. & Ervin, G.N. (2019) A global examination on the differential impacts of roadsides on native *vs.* exotic and weedy plant species. *Global Ecology and Conservation* 17: e00555.

Lemke, A., Kowarik, I. & von der Lippe, M. (2019) How traffic facilitates population expansion of invasive species along roads: the case of common ragweed in Germany. *Journal of Applied Ecology* 56: 413–422.

Lonsdale, W.M. & Lane, A.M. (1994) Tourist vehicles as vectors of weed seeds in Kakadu National Park, Northern Australia. *Biological Conservation* 69: 277–283.

Mortensen, D.A., Rauschert, E.S.J., Nord, A.N. & Jones, B.P. (2009) The role of roads in plant invasions. *Invasive Plant Science and Management* 2: 191–199.

Numminen, E. & Laine, A.-L. (2020) The spread of a wild plant pathogen is driven by the road network. *PLOS Computational Biology* 16: e1007703.

Phillips, B.L., Brown, G.P., Webb, J.K. & Shine, R. (2006) Invasion and the evolution of speed in toads. *Nature* 439: 803.

Raiter, K.G., Hobbs, R.J., Possingham, H.P., Valentine, L.E. & Prober, S.M. (2018) Vehicle tracks are predator highways in intact landscapes. *Biological Conservation* 228: 281–290.

Schmidt, W. (1989) Plant dispersal by motor cars. *Vegetatio* 80: 147–152.

Seabrook, W.A. & Dettmann, E.B. (1996) Roads as activity corridors for cane toads in Australia. *Journal of Wildlife Management* 60: 363–368.

Sharma, G.P. & Raghubanshi, A.S. (2009) Plant invasions along roads: a case study from central highlands, India. *Environmental Monitoring and Assessment* 157: 191–198.

Taylor, K., Brummer, T., Taper, M.L., Wing, A. & Rew, L.J. (2012) Human-mediated long-distance dispersal: an empirical evaluation of seed dispersal by vehicles. *Diversity and Distributions* 18: 942–951.

Urban, M.C. (2006) Road facilitation of trematode infections in snails of northern Alaska. *Conservation Biology* 20: 1143–1149.

Urban, M.C., Phillips, B.L., Skelly, D.K. & Shine, R. (2008) A toad more traveled: the heterogeneous invasion dynamics of cane toads in Australia. *American Naturalist* 171: E134–148.

Von Der Lippe, M. & Kowarik, I. (2007) Long-distance dispersal of plants by vehicles as a driver of plant invasions. *Conservation Biology* 21: 986–996.

von der Lippe, M. & Kowarik, I. (2012) Interactions between propagule pressure and seed traits shape human-mediated seed dispersal along roads. *Perspectives in Plant Ecology, Evolution and Systematics* 14: 123–130.

von der Lippe, M., Bullock, J.M., Kowarik, I., Knopp, T. & Wichmann, M. (2013) Human-mediated dispersal of seeds by the airflow of vehicles. *PLOS ONE* 8: e52733.

Zwaenepoel, A., Roovers, P. & Hermy, M. (2006) Motor vehicles as vectors of plant species from road verges in a suburban environment. *Basic and Applied Ecology* 7: 83–93.

Roadless areas and future development

Alamgir, M., Campbell, M.J., Sloan, S., Goosem, M., Clements, G.R., Mahmoud, M.I. & Laurance, W.F. (2017) Economic, socio-political and environmental risks of road development in the tropics. *Current Biology* 27: R1130-R1140.

Andrade, M.B.T., Ferrante, L. & Fearnside, P.M. (2021) Brazil's Highway BR-319 demonstrates a crucial lack of environmental governance in Amazonia. *Environmental Conservation*, 10.1017/S0376892921000084.

Barber, C.P., Cochrane, M.A., Souza, C.M. & Laurance, W.F. (2014) Roads, deforestation, and the mitigating effect of protected areas in the Amazon. *Biological Conservation* 177: 203–209.

Blake, S., Deem, S.L., Strindberg, S., Maisels, F., Momont, L., Isia, I.-B., Douglas-Hamilton, I., Karesh, W.B. & Kock, M.D. (2008) Roadless wilderness area determines forest elephant movements in the Congo Basin. *PLOS ONE* 3: e3546.

Carter, N., Killion, A., Easter, T., Brandt, J. & Ford, A. (2020) Road development in Asia: assessing the range-wide risks to tigers. *Science Advances* 6: eaaz9619.

Casella, J. & Paranhos Filho, A. (2013) The influence of highway BR262 on the loss of cerrado vegetation cover in southwestern Brazil. *Oecologia Australis* 17: 77–85.

Clements, G.R., Lynam, A.J., Gaveau, D., Yap, W.L., Lhota, S., Goosem, M., Laurance, S. & Laurance, W.F. (2014) Where and how are roads endangering mammals in southeast Asia's forests? *PLOS ONE* 9: e115376.

DellaSala, D.A., Karr, J.R. & Olson, D.M. (2011) Roadless areas and clean water. *Journal of Soil and Water Conservation* 66: 78A.

Dietz, M.S., Barnett, K., Belote, R.T. & Aplet, G.H. (2021) The importance of U.S. national forest roadless areas for vulnerable wildlife species. *Global Ecology and Conservation* 32: e01943.

Dulac, J. (2013) *Global Land Transport Infrastructure Requirements: Estimating Road and Railway Infrastructure Capacity and Costs to 2050.* Paris: International Energy Agency.

Espinosa, S., Celis, G. & Branch, L.C. (2018) When roads appear jaguars decline: increased access to an Amazonian wilderness area reduces potential for jaguar conservation. *PLOS ONE* 13: e0189740.

Fearnside, P.F. (2006) Containing destruction from Brazil's Amazon highways: now is the time to give weight to the environment in decision-making. *Environmental Conservation* 33: 181–183.

Fearnside, P.M. (2007) Brazil's Cuiabá–Santarém (BR-163) Highway: the environmental cost of paving a soybean corridor through the Amazon. *Environmental Management* 39: 601–614.

Ferrante, L., Andrade, M.B.T. & Fearnside, P.M. (2021) Land grabbing on Brazil's Highway BR-319 as a spearhead for Amazonian deforestation. *Land Use Policy* 108: 105559.

Gallice, G.R., Larrea-Gallegos, G. & Vázquez-Rowe, I. (2017) The threat of road expansion in the Peruvian Amazon. *Oryx* 53: 284–292.

Ibisch, P.L., Hoffmann, M.T., Kreft, S., Pe'er, G., Kati, V., Biber-Freudenberger, L., DellaSala, D.A., Vale, M.M., Hobson, P.R. & Selva, N. (2016) A global map of roadless areas and their conservation status. *Science* 354: 1423.

Keshkamat, S.S., Tsendbazar, N.E., Zuidgeest, M.H.P., van der Veen, A. & de Leeuw, J. (2012) The environmental impact of not having paved roads in arid regions: an example from Mongolia. *AMBIO* 41: 202–205.

Laurance, W.F., Albernaz, A.K.M., Schroth, G., Fearnside, P.M., Bergen, S., Venticinque, E.M. & Da Costa, C. (2002) Predictors of deforestation in the Brazilian Amazon. *Journal of Biogeography* 29: 737–748.

Laurance, W.F., Croes, B.M., Tchignoumba, L., Lahm, S.A., Alonso, A., Lee, M.E., Campbell, P. & Ondzeano, C. (2006) Impacts of roads and hunting on central African rainforest mammals. *Conservation Biology* 20: 1251–1261.

Laurance, W.F., Goosem, M. & Laurance, S.G.W. (2009) Impacts of roads and linear clearings on tropical forests. *Trends in Ecology & Evolution* 24: 659–669.

Laurance, W.F., Clements, G.R., Sloan, S., O'Connell, C.S., Mueller, N.D., Goosem, M., Venter, O., Edwards, D.P., Phalan, B., Balmford, A., Van Der Ree, R. & Arrea, I.B. (2014) A global strategy for road building. *Nature* 513: 229–232.

Laurance, W.F. & Burgués Arrea, I. (2017) Roads to riches or ruin? *Science* 358: 442–444.

Laurance, W.F., Campbell, M.J., Alamgir, M. & Mahmoud, M.I. (2017) Road expansion and the fate of Africa's tropical forests. *Frontiers in Ecology and Evolution* 5.

Lupinetti-Cunha, A., Cirino, D.W., Vale, M.M. & Freitas, S.R. (2022) Roadless areas in Brazil: land cover, land use, and conservation status. *Regional Environmental Change* 22: 96.

Meijer, J.R., Huijbregts, M.A.J., Schotten, K.C.G.J. & Schipper, A.M. (2018) Global patterns of current and future road infrastructure. *Environmental Research Letters* 13: 064006.

Mota, N.M., Gastauer, M., Carrión, J.F. & Meira-Neto, J.A.A. (2022) Roads as conduits of functional and phylogenetic degradation in Caatinga. *Tropical Ecology:* 10.1007/s42965-022-00245-x.

Pfaff, A., Robalino, J., Walker, R., Aldrich, S., Caldas, M., Reis, E., Perz, S., Bohrer, C., Arima, E., Laurance, W. & Kirby, K. (2007) Road investments, spatial spillovers, and deforestation in the Brazilian Amazon. *Journal of Regional Science* 47: 109–123.

Psaralexi, M.K., Votsi, N.-E.P., Selva, N., Mazaris, A.D. & Pantis, J.D. (2017) Importance of roadless areas for the European conservation network. *Frontiers in Ecology and Evolution* 5: 2.

Riitters, K.H. & Wickham, J.D. (2003) How far to the nearest road? *Frontiers in Ecology and the Environment* 1: 125–129.

Selva, N., Kreft, S., Kati, V., Schluck, M., Jonsson, B.G., Mihok, B., Okarma, H. & Ibisch, P.L. (2011) Roadless and low-traffic areas as conservation targets in Europe. *Environmental Management* 48: 865–877.

Talty, M.J., Mott Lacroix, K., Aplet, G.H. & Belote, R.T. (2020) Conservation value of national forest roadless areas. *Conservation Science and Practice* 2: e288.

Tisler, T.R., Teixeira, F.Z. & Nóbrega, R.A.A. (2022) Conservation opportunities and challenges in Brazil's roadless and railroad-less areas. *Science Advances* 8: eabi5548.

Vilela, T., Malky Harb, A., Bruner, A., Laísa da Silva Arruda, V., Ribeiro, V., Auxiliadora Costa Alencar, A., Julissa Escobedo Grandez, A., Rojas, A., Laina, A. & Botero, R. (2020) A better Amazon road network for people and the environment. *Proceedings of the National Academy of Sciences* 117: 7095.

Watts, R., Compton, R., McCammon, J., Rich, C., Wright, S., Owens, T. & Ouren, D. (2007) Roadless space of the conterminous United States. *Science* 316: 736–738.

Wilkie, D., Shaw, E., Rotberg, F., Morelli, G. & Auzel, P. (2000) Roads, development, and conservation in the Congo Basin. *Conservation Biology* 14: 1614–1622.

Yang, H., Simmons, B.A., Ray, R., Nolte, C., Gopal, S., Ma, Y., Ma, X. & Gallagher, K.P. (2021) Risks to global biodiversity and Indigenous lands from China's overseas development finance. *Nature Ecology & Evolution*, 10.1038/s41559-021-01541-w.

Zhang, D., Wu, L., Huang, S., Zhang, Z., Ahmad, F., Zhang, G., Shi, N. & Xu, H. (2021) Ecology and environment of the Belt and Road under global climate change: a systematic review of spatial patterns, cost efficiency, and ecological footprints. *Ecological Indicators* 131: 108237.

Chapter 6: Thunder Road
Human health impacts of traffic noise

Andersen, Z.J., Jørgensen, J.T., Elsborg, L., Lophaven, S.N., Backalarz, C., Laursen, J.E., Pedersen, T.H., Simonsen, M.K., Bräuner, E.V. & Lynge, E. (2018) Long-term exposure to road traffic noise and incidence of breast cancer: a cohort study. *Breast Cancer Research* 20: 119.

Cantuaria, M.L., Waldorff, F.B., Wermuth, L., Pedersen, E.R., Poulsen, A.H., Thacher, J.D., Raaschou-Nielsen, O., Ketzel, M., Khan, J., Valencia, V.H., Schmidt, J.H. & Sørensen, M. (2021) Residential exposure to transportation noise in Denmark and incidence of dementia: national cohort study. *BMJ* 374: n1954.

Clark, C., Sbihi, H., Tamburic, L., Brauer, M., Frank, L.D. & Davies, H.W. (2017) Association of long-term exposure to transportation noise and traffic-related air pollution with the incidence of diabetes: a prospective cohort study. *Environmental Health Perspectives* 125: 087025–087025.

Cole-Hunter, T., So, R., Amini, H., Backalarz, C., Brandt, J., Bräuner, E.V., Hertel, O., Jensen, S.S., Jørgensen, J.T., Ketzel, M., Laursen, J.E., Lim, Y.-H., Loft, S., Mehta, A., Mortensen, L.H., Simonsen, M.K., Sisgaard, T., Westendorp, R. & Andersen, Z.J. (2022) Long-term exposure to road traffic noise and all-cause and cause-specific mortality: a Danish nurse cohort study. *Science of The Total Environment*: 153057.

De Coensel, B., Vanwetswinkel, S. & Botteldooren, D. (2011) Effects of natural sounds on the perception of road traffic noise. *Journal of the Acoustical Society of America* 129: EL148-EL153.

Dreger, S., Schüle, S.A., Hilz, L.K. & Bolte, G. (2019) Social inequalities in environmental noise exposure: a review of evidence in the WHO European Region. *International Journal of Environmental Research and Public Health* 16: 1011.

European Environment Agency (2016) *Quiet Areas in Europe: The Environment Unaffected by Noise Pollution.* Copenhagen: EEA.

European Environment Agency (2020) *Environmental Noise in Europe — 2020.* Copenhagen, Denmark: EEA.

Fink, D. (2019) A new definition of noise: noise is unwanted and/or harmful sound. Noise is the new 'secondhand smoke'. *Proceedings of Meetings on Acoustics* 39: 050002.

Foraster, M., Esnaola, M., López-Vicente, M., Rivas, I., Álvarez-Pedrerol, M., Persavento, C., Sebastian-Galles, N., Pujol, J., Dadvand, P. & Sunyer, J. (2022) Exposure to road traffic noise and cognitive development in schoolchildren in Barcelona, Spain: a population-based cohort study. *PLOS Medicine* 19: e1004001.

Goines, L. & Hagler, L. (2007) Noise pollution: a modern plague. *Southern Medical Journal* 100: 287–294.

Hahad, O., Frenis, K., Kuntic, M., Daiber, A. & Münzel, T. (2021) Accelerated aging and age-related diseases (CVD and neurological) due to air pollution and traffic noise exposure. *International Journal of Molecular Sciences* 22.

Halonen, J.I., Hansell, A.L., Gulliver, J., Morley, D., Blangiardo, M., Fecht, D., Toledano, M.B., Beevers, S.D., Anderson, H.R., Kelly, F.J. & Tonne, C. (2015) Road traffic noise is associated with increased cardiovascular morbidity and mortality and all-cause mortality in London. *European Heart Journal* 36: 2653–2661.

Hansell, A., Cai, Y.S. & Gulliver, J. (2017) Cardiovascular health effects of road traffic noise. In: *Environmental Impacts of Road Vehicles: Past, Present and Future*: pp. 107–132. London: The Royal Society of Chemistry.

Hao, G., Zuo, L., Weng, X., Fei, Q., Zhang, Z., Chen, L., Wang, Z. & Jing, C. (2022) Associations of road traffic noise with cardiovascular diseases and mortality: longitudinal results from UK Biobank and meta-analysis. *Environmental Research* 212: 113129.

Hao, Y., Kang, J. & Wortche, H. (2016) Assessment of the masking effects of birdsong on the road traffic noise environment. *Journal of the Acoustical Society of America* 140: 978–987.

Hegewald, J., Schubert, M., Freiberg, A., Romero Starke, K., Augustin, F., Riedel-Heller, S.G., Zeeb, H. & Seidler, A. (2020) Traffic noise and mental health: a systematic review and meta-analysis. *International Journal of Environmental Research and Public Health* 17: 6175.

Khomenko, S., Cirach, M., Barrera-Gómez, J., Pereira-Barboza, E., Iungman, T., Mueller, N., Foraster, M., Tonne, C., Thondoo, M., Jephcote, C., Gulliver, J., Woodcock, J. & Nieuwenhuijsen, M. (2022) Impact of road traffic noise on annoyance and preventable mortality in European cities: a health impact assessment. *Environment International* 162: 107160.

Mac Domhnaill, C., Douglas, O., Lyons, S., Murphy, E. & Nolan, A. (2021) Road traffic noise and cognitive function in older adults: a cross-sectional investigation of the Irish Longitudinal Study on Ageing. *BMC Public Health* 21: 1814.

Moreyra, A.E., Subramanian, K., Mi, Z., Cosgrove, N.M., Kostis, W.J., Ke, F., Kostis, J.B. & Ananth, C. (2022) The impact of exposure to transportation noise on the rates of myocardial infarction in New Jersey. *Journal of the American College of Cardiology* 79: 1148.

Münzel, T., Kröller-Schön, S., Oelze, M., Gori, T., Schmidt, F.P., Steven, S., Hahad, O., Röösli, M., Wunderli, J.-M., Daiber, A. & Sørensen, M. (2020) Adverse cardiovascular effects of traffic noise with a focus on nighttime noise and the new WHO Noise Guidelines. *Annual Review of Public Health* 41: 309–328.

Rossi, I.A., Vienneau, D., Ragettli, M.S., Flückiger, B. & Röösli, M. (2020) Estimating the health benefits associated with a speed limit reduction to thirty kilometres per hour: a health impact assessment of noise and road traffic crashes for the Swiss city of Lausanne. *Environment International* 145: 106126.

Sanok, S., Berger, M., Müller, U., Schmid, M., Weidenfeld, S., Elmenhorst, E.-M. & Aeschbach, D. (2021) Road traffic noise impacts sleep continuity in suburban residents: exposure-response quantification of noise-induced awakenings from vehicle pass-bys at night. *Science of The Total Environment*: 152594.

Shepherd, D., Welch, D., Dirks, K.N. & McBride, D. (2013) Do quiet areas afford greater health-related quality of life than noisy areas? *International Journal of Environmental Research and Public Health* 10.

Smith, R.B., Beevers, S.D., Gulliver, J., Dajnak, D., Fecht, D., Blangiardo, M., Douglass, M., Hansell, A.L., Anderson, H.R., Kelly, F.J. & Toledano, M.B. (2020) Impacts of air pollution and noise on risk of preterm birth and stillbirth in London. *Environment International* 134: 105290.

Sørensen, M., Andersen, Z.J., Nordsborg, R.B., Jensen, S.S., Lillelund, K.G., Beelen, R., Schmidt, E.B., Tjønneland, A., Overvad, K. & Raaschou-Nielsen, O. (2012) Road traffic noise and incident myocardial infarction: a prospective cohort study. *PLOS ONE* 7: e39283.

Swinburn, T.K., Hammer, M.S. & Neitzel, R.L. (2015) Valuing quiet: an economic assessment of U.S. environmental noise as a cardiovascular health hazard. *American Journal of Preventive Medicine* 49: 345–353.

Tangermann, L., Vienneau, D., Hattendorf, J., Saucy, A., Künzli, N., Schäffer, B., Wunderli, J.M. & Röösli, M. (2022) The association of road traffic noise with problem behaviour in adolescents: a cohort study. *Environmental Research*: 112645.

Thacher, J.D., Poulsen, A.H., Raaschou-Nielsen, O., Hvidtfeldt, U.A., Brandt, J., Christensen, J.H., Khan, J., Levin, G., Münzel, T. & Sørensen, M. (2022) Exposure to transportation noise and risk for cardiovascular disease in a nationwide cohort study from Denmark. *Environmental Research* 211: 113106.

Thompson, R., Smith, R.B., Bou Karim, Y., Shen, C., Drummond, K., Teng, C. & Toledano, M.B. (2022) Noise pollution and human cognition: an updated systematic review and meta-analysis of recent evidence. *Environment International* 158: 106905.

van Kempen, E., Casas, M., Pershagen, G. & Foraster, M. (2018) WHO environmental noise guidelines for the European region: a systematic review on environmental noise and cardiovascular and metabolic effects: a summary. *IJERPH* 15: 379.

Vienneau, D., Saucy, A., Schäffer, B., Flückiger, B., Tangermann, L., Stafoggia, M., Wunderli, J.M. & Röösli, M. (2022) Transportation noise exposure and cardiovascular mortality: 15-years of follow-up in a nationwide prospective cohort in Switzerland. *Environment International* 158: 106974.

World Health Organisation (2011) *Burden of Disease from Environmental Noise: Quantification of Healthy Life Years Lost in Europe*. Copenhagen: WHO.

World Health Organisation (2018) *Environmental Noise Guidelines for the European Region*. Copenhagen: WHO Regional Office for Europe.

Impacts of traffic noise on wildlife health

Allen, L.C., Hristov, N.I., Rubin, J.J., Lightsey, J.T. & Barber, J.R. (2021) Noise distracts foraging bats. *Proceedings of the Royal Society B: Biological Sciences* 288: 20202689.

Bonsen, G., Law, B. & Ramp, D. (2015) Foraging strategies determine the effect of traffic noise on bats. *Acta Chiropterologica* 17: 347–357, 11.

Brumm, H., Goymann, W., Derégnaucourt, S., Geberzahn, N. & Zollinger, S.A. (2021) Traffic noise disrupts vocal development and suppresses immune function. *Science Advances* 7: eabe2405.

Castaneda, E., Leavings, V.R., Noss, R.F. & Grace, M.K. (2020) The effects of traffic noise on tadpole behavior and development. *Urban Ecosystems* 23: 245–253.

Chen, H.L. & Koprowski, J.L. (2015) Animal occurrence and space use change in the landscape of anthropogenic noise. *Biological Conservation* 192: 315–322.

Crino, O.L., Van Oorschot, B.K., Johnson, E.E., Malisch, J.L. & Breuner, C.W. (2011) Proximity to a high traffic road: glucocorticoid and life history consequences for nestling white-crowned sparrows. *General and Comparative Endocrinology* 173: 323–332.

Di, G.Q. & Qin, Z.Q. (2018) Influences of combined traffic noise on the ability of learning and memory in mice. *Noise Health* 20: 9–15.

Dietz, M.S., Murdock, C.C., Romero, L.M., Ozgul, A. & Foufopoulos, J. (2013) Distance to a road is associated with reproductive success and physiological stress response in a migratory landbird. *Wilson Journal of Ornithology* 125: 50–61.

Dorado-Correa, A.M., Zollinger, S.A., Heidinger, B. & Brumm, H. (2018) Timing matters: traffic noise accelerates telomere loss rate differently across developmental stages. *Frontiers in Zoology* 15: 29.

Duque, F.G., Rodriguez-Saltos, C.A., Uma, S., Nasir, I., Monteros, M.F., Wilczynski, W. & Carruth, L.L. (2020) High-frequency hearing in a hummingbird. *Science Advances* 6: eabb9393.

Finch, D., Schofield, H. & Mathews, F. (2020) Traffic noise playback reduces the activity and feeding behaviour of free-living bats. *Environmental Pollution* 263: 114405.

Flores, R., Penna, M., Wingfield, J.C., Cuevas, E., Vásquez, R.A. & Quirici, V. (2020) Effects of traffic noise exposure on corticosterone, glutathione and tonic immobility in chicks of a precocial bird. *Conservation Physiology* 7.

Geerts, S. & Pauw, A. (2011) Easy technique for assessing pollination rates in the genus *Erica* reveals road impact on bird pollination in the Cape fynbos, South Africa. *Austral Ecology* 36: 656–662.

Grunst, A.S., Grunst, M.L., Bervoets, L., Pinxten, R. & Eens, M. (2020a) Proximity to roads, but not exposure to metal pollution, is associated with accelerated developmental telomere shortening in nestling great tits. *Environmental Pollution* 256: 113373.

Grunst, M.L., Grunst, A.S., Pinxten, R. & Eens, M. (2020b) Anthropogenic noise is associated with telomere length and carotenoid-based coloration in free-living nestling songbirds. *Environmental Pollution* 260: 114032.

Grunst, M.L., Grunst, A.S., Pinxten, R. & Eens, M. (2021) Variable and consistent traffic noise negatively affect the sleep behavior of a free-living songbird. *Science of The Total Environment* 778: 146338.

Halfwerk, W., Holleman, L.J.M., Lessells, C.M. & Slabbekoorn, H. (2011) Negative impact of traffic noise on avian reproductive success. *Journal of Applied Ecology* 48: 210–219.

Harding, H.R., Gordon, T.A.C., Eastcott, E., Simpson, S.D. & Radford, A.N. (2019) Causes and consequences of intraspecific variation in animal responses to anthropogenic noise. *Behavioral Ecology* 30: 1501–1511.

Hayward, L.S., Bowles, A.E., Ha, J.C. & Wasser, S.K. (2011) Impacts of acute and long-term vehicle exposure on physiology and reproductive success of the northern spotted owl. *Ecosphere* 2: art65.

Ibáñez-Álamo, J.D., Pineda-Pampliega, J., Thomson, R.L., Aguirre, J.I., Díez-Fernández, A., Faivre, B., Figuerola, J. & Verhulst, S. (2018) Urban blackbirds have shorter telomeres. *Biology Letters* 14: 20180083.

Iglesias-Merchan, C., Horcajada-Sánchez, F., Diaz-Balteiro, L., Escribano-Ávila, G., Lara-Romero, C., Virgós, E., Planillo, A. & Barja, I. (2018) A new large-scale index (AcED) for assessing traffic noise disturbance on wildlife: stress response in a roe deer (*Capreolus capreolus*) population. *Environmental Monitoring and Assessment* 190: 185.

Injaian, A.S., Poon, L.Y. & Patricelli, G.L. (2018a) Effects of experimental anthropogenic noise on avian settlement patterns and reproductive success. *Behavioral Ecology* 29: 1181–1189.

Injaian, A.S., Taff, C.C. & Patricelli, G.L. (2018b) Experimental anthropogenic noise impacts avian parental behaviour, nestling growth and nestling oxidative stress. *Animal Behaviour* 136: 31–39.

Injaian, A.S., Taff, C.C., Pearson, K.L., Gin, M.M.Y., Patricelli, G.L. & Vitousek, M.N. (2018c) Effects of experimental chronic traffic noise exposure on adult and nestling corticosterone levels, and nestling body condition in a free-living bird. *Hormones and Behavior* 106: 19–27.

Injaian, A.S., Gonzalez-Gomez, P.L., Taff, C.C., Bird, A.K., Ziur, A.D., Patricelli, G.L., Haussmann, M.F. & Wingfield, J.C. (2019) Traffic noise exposure alters nestling physiology and telomere attrition through direct, but not maternal, effects in a free-living bird. *General and Comparative Endocrinology* 276: 14–21.

Jafari, Z., Kolb, B.E. & Mohajerani, M.H. (2018) Chronic traffic noise stress accelerates brain impairment and cognitive decline in mice. *Experimental Neurology* 308: 1–12.

Kafash, Z.H., Khoramnejadian, S., Ghotbi-Ravandi, A.A. & Dehghan, S.F. (2022) Traffic noise induces oxidative stress and phytohormone imbalance in two urban plant species. *Basic and Applied Ecology*.

Kight, C.R., Saha, M.S. & Swaddle, J.P. (2012) Anthropogenic noise is associated with reductions in the productivity of breeding Eastern Bluebirds (*Sialia sialis*). *Ecological Applications* 22: 1989–96.

Kleist, N.J., Guralnick, R.P., Cruz, A., Lowry, C.A. & Francis, C.D. (2018) Chronic anthropogenic noise disrupts glucocorticoid signaling and has multiple effects on fitness in an avian community. *Proceedings of the National Academy of Sciences* 115: E648.

Kunc, H.P. & Schmidt, R. (2019) The effects of anthropogenic noise on animals: a meta-analysis. *Biology Letters* 15: 20190649.

Lucass, C., Eens, M. & Müller, W. (2016) When ambient noise impairs parent–offspring communication. *Environ Pollut* 212: 592–597.

Lunde, E.T., Bech, C., Fyumagwa, R.D., Jackson, C.R. & Røskaft, E. (2016) Assessing the effect of roads on impala (*Aepyceros melampus*) stress levels using faecal glucocorticoid metabolites. *African Journal of Ecology* 54: 434–441.

Mason, J.T., McClure, C.J.W. & Barber, J.R. (2016) Anthropogenic noise impairs owl hunting behavior. *Biological Conservation* 199: 29–32.

Meillère, A., Brischoux, F., Ribout, C. & Angelier, F. (2015) Traffic noise exposure affects telomere length in nestling house sparrows. *Biology Letters* 11: 20150559-20150559.

Morley, E.L., Jones, G. & Radford, A.N. (2014) The importance of invertebrates when considering the impacts of anthropogenic noise. *Proceedings of the Royal Society B: Biological Sciences* 281: 20132683.

Ortega, C.P. (2012) Effects of noise pollution on birds: a brief review of our knowledge. *Ornithological Monographs* 74: 6–22.

Osbrink, A., Meatte, M.A., Tran, A., Herranen, K.K., Meek, L., Murakami-Smith, M., Ito, J., Bhadra, S., Nunnenkamp, C. & Templeton, C.N. (2021) Traffic noise inhibits cognitive performance in a songbird. *Proceedings of the Royal Society B: Biological Sciences* 288: 20202851.

Owens, J.L., Stec, C.L. & O'Hatnick, A. (2012) The effects of extended exposure to traffic noise on parid social and risk-taking behavior. *Behavioural Processes* 91: 61–69.

Potvin, D.A. & MacDougall-Shackleton, S.A. (2015) Traffic noise affects embryo mortality and nestling growth rates in captive zebra finches. *Journal of Experimental Zoology Part A Ecological Genetics and Physiology* 323: 722–730.

Ribeiro, P.V.A., Gonçalves, V.F., de Magalhães Tolentino, V.C., Baesse, C.Q., Pires, L.P., Paniago, L.P.M. & de Melo, C. (2022) Effects of urbanisation and pollution on the heterophil/lymphocyte ratio in birds from Brazilian Cerrado. *Environmental Science and Pollution Research*.

Salmón, P., Nilsson, J.F., Nord, A., Bensch, S. & Isaksson, C. (2016) Urban environment shortens telomere length in nestling great tits, *Parus major*. *Biology Letters* 12: 20160155.

Schroeder, J., Nakagawa, S., Cleasby, I.R. & Burke, T. (2012) Passerine birds breeding under chronic noise experience reduced fitness. *PLOS ONE* 7: e39200.

Senzaki, M., Yamaura, Y., Francis, C.D. & Nakamura, F. (2016) Traffic noise reduces foraging efficiency in wild owls. *Scientific Reports* 6: 30602.

Senzaki, M., Barber, J.R., Phillips, J.N., Carter, N.H., Cooper, C.B., Ditmer, M.A., Fristrup, K.M., McClure, C.J.W., Mennitt, D.J., Tyrrell, L.P., Vukomanovic, J., Wilson, A.A. & Francis, C.D. (2020) Sensory pollutants alter bird phenology and fitness across a continent. *Nature* 587: 605–609.

Song, S., Chang, Y., Wang, D., Jiang, T., Feng, J. & Lin, A. (2020) Chronic traffic noise increases food intake and alters gene expression associated with metabolism and disease in bats. *Journal of Applied Ecology* 57: 1915–1925.

Tennessen, J.B., Parks, S.E. & Langkilde, T. (2014) Traffic noise causes physiological stress and impairs breeding migration behaviour in frogs. *Conservation Physiology* 2: cou032.

Troïanowski, M., Mondy, N., Dumet, A., Arcanjo, C. & Lengagne, T. (2017) Effects of traffic noise on tree frog stress levels, immunity, and color signaling. *Conservation Biology* 31: 1132–1140.

Zollinger, S.A., Dorado-Correa, A., Goymann, W., Forstmeier, W., Knief, U., Bastidas-Urrutia, A.M. & Brumm, H. (2020) Traffic noise exposure depresses plasma corticosterone and delays offspring growth in breeding zebra finches. *Conservation Physiology* 7: 10.1093/conphys/coz056.

Impacts of traffic noise on wildlife communication and behaviour

Arevalo, J. & Blau, E. (2018) Road encroachment near protected areas alters the natural soundscape through traffic noise pollution in Costa Rica. *Revista de CIENCIAS AMBIENTALES, Tropical Journal of Environmental Sciences* 52: 2215–3896.

Barber, J.R., Crooks, K.R. & Fristrup, K.M. (2010) The costs of chronic noise exposure for terrestrial organisms. *Trends in Ecology & Evolution* 25: 180–9.

Barber, J.R., Burdett, C.L., Reed, S.E., Warner, K.A., Formichella, C., Crooks, K.R., Theobald, D.M. & Fristrup, K.M. (2011) Anthropogenic noise exposure in protected natural areas: estimating the scale of ecological consequences. *Landscape Ecology* 26: 1281–1295.

Bednarz, P.A. (2021) Do decibels matter? A review of effects of traffic noise on terrestrial small mammals and bats. *Polish Journal of Ecology* 68: 323–333.

Bee, M.A. & Swanson, E.M. (2007) Auditory masking of anuran advertisement calls by road traffic noise. *Animal Behaviour* 74: 1765–1776.

Bent, A.M., Ings, T.C. & Mowles, S.L. (2021) Anthropogenic noise disrupts mate choice behaviors in female *Gryllus bimaculatus*. *Behavioral Ecology*, 10.1093/beheco/araa124.

Bermúdez-Cuamatzin, E., Ríos-Chelén, A.A., Gil, D. & Garcia, C.M. (2011) Experimental evidence for real-time song frequency shift in response to urban noise in a passerine bird. *Biology Letters* 7: 36–38.

Brumm, H. (2004) The impact of environmental noise on song amplitude in a territorial bird. *Journal of Animal Ecology* 73: 434–440.

Brumm, H., Goymann, W., Derégnaucourt, S., Geberzahn, N. & Zollinger, S.A. (2021) Traffic noise disrupts vocal development and suppresses immune function. *Science Advances* 7: eabe2405.

Buxton, R.T., McKenna, M.F., Mennitt, D., Fristrup, K., Crooks, K., Angeloni, L. & Wittemyer, G. (2017) Noise pollution is pervasive in U.S. protected areas. *Science* 356: 531–533.

Byrnes, P., Goosem, M. & Turton, S.M. (2012) Are less vocal rainforest mammals susceptible to impacts from traffic noise? *Wildlife Research* 39: 355–365.

Caorsi, V.Z., Both, C., Cechin, S., Antunes, R. & Borges-Martins, M. (2017) Effects of traffic noise on the calling behavior of two Neotropical hylid frogs. *PLOS ONE* 12: e0183342.

Classen-Rodríguez, L., Tinghitella, R. & Fowler-Finn, K. (2021) Anthropogenic noise affects insect and arachnid behavior, thus changing interactions within and between species. *Current Opinion in Insect Science* 47: 142–153.

Cunnington, G.M. & Fahrig, L. (2010) Plasticity in the vocalizations of anurans in response to traffic noise. *Acta Oecologica* 36: 463–470.

Cunnington, G.M. & Fahrig, L. (2013) Mate attraction by male anurans in the presence of traffic noise. *Animal Conservation* 16: 275–285.

Damsky, J. & Gall, M.D. (2016) Anthropogenic noise reduces approach of Black-capped Chickadee (*Poecile atricapillus*) and Tufted Titmouse (*Baeolophus bicolor*) to Tufted Titmouse mobbing calls. *The Condor* 119: 26–33.

Derryberry, E.P., Phillips, J.N., Derryberry, G.E., Blum, M.J. & Luther, D. (2020) Singing in a silent spring: birds respond to a half-century soundscape reversion during the COVID-19 shutdown. *Science* 370: 575–579.

Dooling, R.J. & Popper, A.N. (2007) *The Effects of Highway Noise on Birds*. Rockville, MD: Environmental BioAcoustics LLC.

Dooling, R.J. & Popper, A.N. (2016) *Technical Guidance for Assessment and Mitigation of the Effects of Highway and Road Construction Noise on Birds*. Sacramento: California Department of Transportation.

Dooling, R.J., Buehler, R., Leek, M.R. & Popper, A.N. (2019) The impact of urban and traffic noise on birds. *Acoustics Today* 15: 19–27.

Duquette, C.A., Loss, S.R. & Hovick, T.J. (2021) A meta-analysis of the influence of anthropogenic noise on terrestrial wildlife communication strategies. *Journal of Applied Ecology* 58: 1112–1121.

Francis, C.D., Kleist, N.J., Ortega, C.P. & Cruz, A. (2012) Noise pollution alters ecological services: enhanced pollination and disrupted seed dispersal. *Proceedings of the Royal Society B: Biological Sciences* 279: 2727–2735.

Fuller, R.A., Warren, P.H. & Gaston, K.J. (2007) Daytime noise predicts nocturnal singing in urban robins. *Biology Letters* 3: 368–370.

Gentry, K.E., Derryberry, E.P., Danner, R.M., Danner, J.E. & Luther, D.A. (2017) Immediate signaling flexibility in response to experimental noise in urban, but not rural, white-crowned sparrows. *Ecosphere* 8: e01916.

Gentry, K.E., McKenna, M.F. & Luther, D.A. (2018) Evidence of suboscine song plasticity in response to traffic noise fluctuations and temporary road closures. *Bioacoustics* 27: 165–181.

Giordano, A., Hunninck, L. & Sheriff, M.J. (2022) Prey responses to predation risk under chronic road noise. *Journal of Zoology* 317: 147–157.

Grade, A.M. & Sieving, K.E. (2016) When the birds go unheard: highway noise disrupts information transfer between bird species. *Biology Letters* 12: 20160113.

Grobler, B.A. & Campbell, E.E. (2022) Road and landscape-context impacts on bird pollination in fynbos of the southeastern Cape Floristic Region. *South African Journal of Botany* 146: 676–684.

Halfwerk, W., Bot, S., Buikx, J., van der Velde, M., Komdeur, J., ten Cate, C. & Slabbekoorn, H. (2011) Low-frequency songs lose their potency in noisy urban conditions. *Proceedings of the National Academy of Sciences* 108: 14549.

Halfwerk, W., Lea, A.M., Guerra, M.A., Page, R.A. & Ryan, M.J. (2016) Vocal responses to noise reveal the presence of the Lombard effect in a frog. *Behavioral Ecology* 27: 669–676.

Halfwerk, W. & van Oers, K. (2020) Anthropogenic noise impairs foraging for cryptic prey via cross-sensory interference. *Proceedings of the Royal Society B: Biological Sciences* 287: 20192951.

Hanna, D., Blouin-Demers, G., Wilson, D.R. & Mennill, D.J. (2011) Anthropogenic noise affects song structure in red-winged blackbirds (*Agelaius phoeniceus*). *Journal of Experimental Biology* 214: 3549–3556.

Higham, V., Deal, N.D.S., Chan, Y.K., Chanin, C., Davine, E., Gibbings, G., Keating, R., Kennedy, M., Reilly, N., Symons, T., Vran, K. & Chapple, D.G. (2021) Traffic noise drives an immediate increase in call pitch in an urban frog. *Journal of Zoology* 313: 307–315.

Holt, D.E. & Johnston, C.E. (2015) Traffic noise masks acoustic signals of freshwater stream fish. *Biological Conservation* 187: 27–33.

Jiang, T., Guo, X., Lin, A., Hui, W., Sun, C., Feng, J. & Kanwal, J. (2019) Bats increase vocal amplitude and decrease vocal complexity to mitigate noise interference during social communication. *Animal Cognition* 22: 10.1007/s10071-018-01235-0.

Jung, H., Sherrod, A., LeBreux, S., Price, J.M. & Freeberg, T.M. (2020) Traffic noise and responses to a simulated approaching avian predator in mixed-species flocks of chickadees, titmice, and nuthatches. *Ethology* 126: 620–629.

Kaiser, K. & Hammers, J. (2009) The effect of anthropogenic noise on male advertisement call rate in the neotropical treefrog, *Dendropsophus triangulum*. *Behaviour* 146: 1053–1069.

Lampe, U., Schmoll, T., Franzke, A. & Reinhold, K. (2012) Staying tuned: grasshoppers from noisy roadside habitats produce courtship signals with elevated frequency components. *Functional Ecology* 26: 1348–1354.

Laurance, W.F. (2015) Wildlife struggle in an increasingly noisy world. *Proceedings of the National Academy of Sciences of the USA* 112: 11995–6.

Lengagne, T. (2008) Traffic noise affects communication behaviour in a breeding anuran, *Hyla arborea*. *Biological Conservation* 141: 2023–2031.

Lohrey, K. & Talarczyk, P. (2021) The impact of traffic noise pollution on plant growth within urban community gardens. *Journal of Student Research* 10: https://doi.org/10.47611/jsrhs.v10i1.1352.

Luther, D.A., Phillips, J. & Derryberry, E.P. (2016) Not so sexy in the city: urban birds adjust songs to noise but compromise vocal performance. *Behavioral Ecology* 27: 332–340.

McClure, C.J.W., Ware, H.E., Carlisle, J., Kaltenecker, G. & Barber, J.R. (2013) An experimental investigation into the effects of traffic noise on distributions of birds: avoiding the phantom road. *Proceedings of the Royal Society B: Biological Sciences* 280: 20132290.

McClure, C.J.W., Ware, H.E., Carlisle, J.D. & Barber, J.R. (2017) Noise from a phantom road experiment alters the age structure of a community of migrating birds. *Animal Conservation* 20: 164–172.

Parris, K.M. & Schneider, A. (2009) Impacts of traffic noise and traffic volume on birds of roadside habitats. *Ecology and Society* 14: 29.

Parris, K.M., Velik-Lord, M. & North, J.M.A. (2009) Frogs call at a higher pitch in traffic noise. *Ecology and Society* 14: 25.

Phillips, J.N., Termondt, S.E. & Francis, C.D. (2021) Long-term noise pollution affects seedling recruitment and community composition, with negative effects persisting after removal. *Proceedings of the Royal Society B: Biological Sciences* 288: 20202906.

Pijanowski, B.C., Farina, A., Gage, S.H., Dumyahn, S.L. & Krause, B.L. (2011a) What is soundscape ecology? An introduction and overview of an emerging new science. *Landscape Ecology* 26: 1213–1232.

Pijanowski, B.C., Villanueva-Rivera, L.J., Dumyahn, S.L., Farina, A., Krause, B.L., Napoletano, B.M., Gage, S.H. & Pieretti, N. (2011b) Soundscape ecology: the science of sound in the landscape. *BioScience* 61: 203–216.

Polak, M. (2014) Relationship between traffic noise levels and song perch height in a common passerine bird. *Transportation Research Part D: Transport and Environment* 30: 72–75.

Roca, I.T., Desrochers, L., Giacomazzo, M., Bertolo, A., Bolduc, P., Deschesnes, R., Martin, C.A., Rainville, V., Rheault, G. & Proulx, R. (2016) Shifting song frequencies in response to anthropogenic noise: a meta-analysis on birds and anurans. *Behavioral Ecology* 27: 1269–1274.

Shannon, G., Angeloni, L.M., Wittemyer, G., Fristrup, K.M. & Crooks, K.R. (2014) Road traffic noise modifies behaviour of a keystone species. *Animal Behaviour* 94: 135–141.

Shannon, G., Crooks, K.R., Wittemyer, G., Fristrup, K.M. & Angeloni, L.M. (2016a) Road noise causes earlier predator detection and flight response in a free-ranging mammal. *Behavioral Ecology* 27: 1370–1375.

Shannon, G., McKenna, M.F., Angeloni, L.M., Crooks, K.R., Fristrup, K.M., Brown, E., Warner, K.A., Nelson, M.D., White, C., Briggs, J., McFarland, S. & Wittemyer, G. (2016b) A synthesis of two decades of research documenting the effects of noise on wildlife. *Biological Reviews* 91: 982–1005.

Shannon, G., McKenna, M., Wilson-Henjum, G., Angeloni, L., Crooks, K. & Wittemyer, G. (2020) Vocal characteristics of prairie dog alarm calls across an urban noise gradient. *Behavioral Ecology*.

Shier, D., Lea, A. & Owen, M. (2012) Beyond masking: Endangered Stephens' kangaroo rats respond to traffic noise with footdrumming. *Biological Conservation* 150: 53–58.

Siemers, B.M. & Schaub, A. (2011) Hunting at the highway: traffic noise reduces foraging efficiency in acoustic predators. *Proceedings of the Royal Society B-Biological Sciences* 278: 1646–1652.

Slabbekoorn, H. & Peet, M. (2003) Birds sing at a higher pitch in urban noise. *Nature* 424: 267.

Song, S., Lin, A., Jiang, T., Zhao, X., Metzner, W. & Feng, J. (2019) Bats adjust temporal parameters of echolocation pulses but not those of communication calls in response to traffic noise. *Integrative Zoology* 14: 576–588.

Sweet, K.A., Sweet, B.P., Gomes, D.G.E., Francis, C.D. & Barber, J.R. (2021) Natural and anthropogenic noise increase vigilance and decrease foraging behaviors in song sparrows. *Behavioral Ecology*: arab141.

Templeton, C.N., Zollinger, S.A. & Brumm, H. (2016) Traffic noise drowns out great tit alarm calls. *Current Biology* 26: R1173–1174.

Tolentino, V.C.d.M., Baesse, C.Q. & Melo, C.d. (2018) Dominant frequency of songs in tropical bird species is higher in sites with high noise pollution. *Environmental Pollution* 235: 983–992.

Troïanowski, M., Condette, C., Mondy, N., Dumet, A. & Lengagne, T. (2015) Traffic noise affects colouration but not calls in the European treefrog (*Hyla arborea*). *Behaviour* 152: 821–836.

Zhao, L., Sun, X., Chen, Q., Yang, Y., Wang, J., Ran, J., Brauth, S.E., Tang, Y. & Cui, J. (2018) Males increase call frequency, not intensity, in response to noise, revealing no Lombard effect in the little torrent frog. *Ecology and Evolution* 8: 11733–11741.

Zollinger, S.A., Slater, P.J.B., Nemeth, E. & Brumm, H. (2017) Higher songs of city birds may not be an individual response to noise. *Proceedings of the Royal Society B: Biological Sciences* 284: 20170602.

Chapter 7: Emission Creep

Gaseous and particulate pollution from traffic

Air Quality Expert Group (2019) *Non-Exhaust Emissions from Road Traffic*. London: Department for Environment, Food and Rural Affairs; Scottish Government; Welsh Government; and Department of the Environment in Northern Ireland.

Adamiec, E., Jarosz-Krzemińska, E. & Wieszała, R. (2016) Heavy metals from non-exhaust vehicle emissions in urban and motorway road dusts. *Environmental Monitoring and Assessment* 188: 369–369.

Amato, F., Cassee, F.R., Denier van der Gon, H.A., Gehrig, R., Gustafsson, M., Hafner, W., Harrison, R.M., Jozwicka, M., Kelly, F.J., Moreno, T., Prevot, A.S., Schaap, M., Sunyer, J. & Querol, X. (2014) Urban air quality: the challenge of traffic non-exhaust emissions. *Journal of Hazardous Materials* 275: 31–36.

Andersson-Sköld, Y., Johannesson, M., Gustafsson, M., Järlskog, I., Lithner, D., Polukarova, M. & Strömvall, A.-M. (2020) *Microplastics from Tyre and Road Wear: A Literature Review.* Swedish National Road and Transport Research Institute: VTI rapport 1028A.

Baensch-Baltruschat, B., Kocher, B., Stock, F. & Reifferscheid, G. (2020) Tyre and road wear particles (TRWP) - A review of generation, properties, emissions, human health risk, ecotoxicity, and fate in the environment. *Science of The Total Environment* 733: 137823.

Beevers, S.D., Westmoreland, E., de Jong, M.C., Williams, M.L. & Carslaw, D.C. (2012) Trends in NOx and NO_2 emissions from road traffic in Great Britain. *Atmospheric Environment* 54: 107–116.

Bergmann, M., Mützel, S., Primpke, S., Tekman Mine, B., Trachsel, J. & Gerdts, G. (2021) White and wonderful? Microplastics prevail in snow from the Alps to the Arctic. *Science Advances* 5: eaax1157.

Browne, M.A., Niven, S.J., Galloway, T.S., Rowland, S.J. & Thompson, R.C. (2013) Microplastic moves pollutants and additives to worms, reducing functions linked to health and biodiversity. *Current Biology* 23: 2388–2392.

Evangeliou, N., Grythe, H., Klimont, Z., Heyes, C., Eckhardt, S., Lopez-Aparicio, S. & Stohl, A. (2020) Atmospheric transport is a major pathway of microplastics to remote regions. *Nature Communications* 11: 3381.

Greenfield, D., Sterry, M., Ibbott, N. & Greenfield, J. (2021) *Invisible Ocean Pollutants from our Roads. SUEZ 2030 Futures Insight Report no. 1.* UK: Social, Environmental & Economic Solutions (SOENECS) Ltd.

Grigoratos, T. & Martini, G. (2015) Brake wear particle emissions: a review. *Environmental Science and Pollution Research International* 22: 2491–504.

Khare, P., Machesky, J., Soto, R., He, M., Presto, A.A. & Gentner, D.R. (2020) Asphalt-related emissions are a major missing nontraditional source of secondary organic aerosol precursors. *Science Advances* 6: eabb9785.

Knight, L.J., Parker-Jurd, F.N.F., Al-Sid-Cheikh, M. & Thompson, R.C. (2020) Tyre wear particles: an abundant yet widely unreported microplastic? *Environmental Science and Pollution Research International* 27: 18345–18354.

Kole, P.J., Löhr, A.J., Van Belleghem, F.G.A.J. & Ragas, A.M.J. (2017) Wear and tear of tyres: a stealthy source of microplastics in the environment. *International Journal of Environmental Research and Public Health* 14: 1265.

Leonard, R.J. & Hochuli, D.F. (2017) Exhausting all avenues: why impacts of air pollution should be part of road ecology. *Frontiers in Ecology and the Environment* 15: 443–449.

Materić, D., Kjær, H.A., Vallelonga, P., Tison, J.-L., Röckmann, T. & Holzinger, R. (2022) Nanoplastics measurements in northern and southern polar ice. *Environmental Research* 208: 112741.

Migaszewski, Z., Gałuszka, A., Dołęgowska, S. & Michalik, A. (2021) Glass microspheres in road dust of the city of Kielce (south-central Poland) as markers of traffic-related pollution. *Journal of Hazardous Materials* 413: 125355.

Panko, J.M., Kreider, M.L., McAtee, B.L. & Marwood, C. (2013b) Chronic toxicity of tire and road wear particles to water- and sediment-dwelling organisms. *Ecotoxicology* 22: 13–21.

Panko, J.M., Hitchcock, K.M., Fuller, G.W. & Green, D. (2019) Evaluation of tire wear contribution to PM2.5 in urban environments. *Atmosphere,* 10, 10.3390/atmos10020099.

Perugini, M., Manera, M., Grotta, L., Abete, M.C., Tarasco, R. & Amorena, M. (2011) Heavy metal (Hg, Cr, Cd, and Pb) contamination in urban areas and wildlife reserves: honeybees as bioindicators. *Biological Trace Element Research* 140: 170–176.

Phillips, B.B., Bullock, J.M., Osborne, J.L. & Gaston, K.J. (2021) Spatial extent of road pollution: A national analysis. *Science of The Total Environment* 773: 145589.

Polukarova, M., Markiewicz, A., Björklund, K., Strömvall, A.-M., Galfi, H., Andersson Sköld, Y., Gustafsson, M., Järlskog, I. & Aronsson, M. (2020) Organic pollutants, nano- and microparticles in street sweeping road dust and washwater. *Environment International* 135: 105337.

Resongles, E., Dietze, V., Green, D.C., Harrison, R.M., Ochoa-Gonzalez, R., Tremper, A.H. & Weiss, D.J. (2021) Strong evidence for the continued contribution of lead deposited during the 20th century to the atmospheric environment in London of today. *Proceedings of the National Academy of Sciences* 118: e2102791118.

Szwalec, A., Mundała, P., Kędzior, R. & Pawlik, J. (2020) Monitoring and assessment of cadmium, lead, zinc and copper concentrations in arable roadside soils in terms of different traffic conditions. *Environmental Monitoring and Assessment* 192: 155.

Thorpe, A. & Harrison, R.M. (2008) Sources and properties of non-exhaust particulate matter from road traffic: a review. *Science of The Total Environment* 400: 270–82.

UK Environment Agency (2018) *The state of the environment: air quality.* Environment Agency.

Wagner, S., Hüffer, T., Klöckner, P., Wehrhahn, M., Hofmann, T. & Reemtsma, T. (2018) Tire wear particles in the aquatic environment – a review on generation, analysis, occurrence, fate and effects. *Water Research* 139: 83–100.

Wang, G., Yan, X., Zhang, F., Zeng, C. & Gao, D. (2013) Traffic-related trace element accumulation in roadside soils and wild grasses in the Qinghai-Tibet Plateau, China. *International Journal of Environmental Research and Public Health* 11: 456–472.

Wright, S.L., Rowe, D., Thompson, R.C. & Galloway, T.S. (2013) Microplastic ingestion decreases energy reserves in marine worms. *Current Biology* 23: R1031-R1033.

Wu, T., Lo, K. & Stafford, J. (2021) Vehicle non-exhaust emissions - revealing the pathways from source to environmental exposure. *Environmental Pollution* 268: 115654.

Zhang, Q., Jiang, X., Tong, D., Davis, S.J., Zhao, H., Geng, G., Feng, T., Zheng, B., Lu, Z., Streets, D.G., Ni, R., Brauer, M., van Donkelaar, A., Martin, R.V., Huo, H., Liu, Z., Pan, D., Kan, H., Yan, Y., Lin, J., He, K. & Guan, D. (2017) Transboundary health impacts of transported global air pollution and international trade. *Nature* 543: 705–709.

Tyre wear and coho salmon

Brinkmann, M., Montgomery, D., Selinger, S., Miller, J.G.P., Stock, E., Alcaraz, A.J., Challis, J.K., Weber, L., Janz, D., Hecker, M. & Wiseman, S. (2022) Acute toxicity of the tire rubber-derived chemical 6PPD-quinone to four fishes of commercial, cultural, and ecological importance. *Environmental Science & Technology Letters* 9: 333–338.

Cao, G., Wang, W., Zhang, J., Wu, P., Zhao, X., Yang, Z., Hu, D. & Cai, Z. (2022) New evidence of rubber-derived quinones in water, air, and soil. *Environmental Science & Technology* 56: 4142–4150.

Chow, M.I., Lundin, J.I., Mitchell, C.J., Davis, J.W., Young, G., Scholz, N.L. & McIntyre, J.K. (2019) An urban stormwater runoff mortality syndrome in juvenile coho salmon. *Aquatic Toxicology* 214: 105231.

Hiki, K., Asahina, K., Kato, K., Yamagishi, T., Omagari, R., Iwasaki, Y., Watanabe, H. & Yamamoto, H. (2021) Acute toxicity of a tire rubber-derived chemical, 6PPD quinone, to freshwater fish and crustacean species. *Environmental Science & Technology Letters* 8: 779–784.

Johannessen, C., Helm, P., Lashuk, B., Yargeau, V. & Metcalfe, C.D. (2022) The tire wear compounds 6PPD-quinone and 1,3-diphenylguanidine in an urban watershed. *Archives of Environmental Contamination and Toxicology* 82: 171–179.

Marwood, C., McAtee, B., Kreider, M., Ogle, R.S., Finley, B., Sweet, L. & Panko, J. (2011) Acute aquatic toxicity of tire and road wear particles to alga, daphnid, and fish. *Ecotoxicology* 20: 2079.

McIntyre, J.K., Prat, J., Cameron, J., Wetzel, J., Mudrock, E., Peter, K.T., Tian, Z., Mackenzie, C., Lundin, J., Stark, J.D., King, K., Davis, J.W., Kolodziej, E.P. & Scholz, N.L. (2021) Treading water: tire wear particle leachate recreates an urban runoff mortality syndrome in coho but not chum salmon. *Environmental Science & Technology* 55: 11767–11774.

Peter, K.T., Tian, Z., Wu, C., Lin, P., White, S., Du, B., McIntyre, J.K., Scholz, N.L. & Kolodziej, E.P. (2018) Using high-resolution mass spectrometry to identify organic contaminants linked to urban stormwater mortality syndrome in coho salmon. *Environmental Science & Technology* 52: 10317–10327.

Rauert, C., Charlton, N., Okoffo, E.D., Stanton, R.S., Agua, A.R., Pirrung, M.C. & Thomas, K.V. (2022) Concentrations of tire additive chemicals and tire road wear particles in an Australian urban tributary. *Environmental Science & Technology* 56: 2421–2431.

Tian, Z., Zhao, H., Peter, K., T., Gonzalez, M., Wetzel, J., Wu, C., Hu, X., Prat, J., Mudrock, E., Hettinger, R., Cortina Allan, E., Biswas Rajshree, G., Kock Flávio Vinicius, C., Soong, R., Jenne, A., Du, B., Hou, F., He, H., Lundeen, R., Gilbreath, A., Sutton, R., Scholz Nathaniel, L., Davis Jay, W., Dodd Michael, C., Simpson, A., McIntyre Jenifer, K. & Kolodziej Edward, P. (2021) A ubiquitous tire rubber–derived chemical induces acute mortality in coho salmon. *Science* 371: 185–189.

Tian, Z., Gonzalez, M., Rideout, C.A., Zhao, H.N., Hu, X., Wetzel, J., Mudrock, E., James, C.A., McIntyre, J.K. & Kolodziej, E.P. (2022) 6PPD-Quinone: revised toxicity assessment and quantification with a commercial standard. *Environmental Science & Technology Letters* 9: 140–146.

Varshney, S., Gora, A.H., Siriyappagouder, P., Kiron, V. & Olsvik, P.A. (2022) Toxicological effects of 6PPD and 6PPD quinone in zebrafish larvae. *Journal of Hazardous Materials* 424: 127623.

Human health impacts of traffic pollution

Avagyan, R., Sadiktsis, I., Bergvall, C. & Westerholm, R. (2014) Tire tread wear particles in ambient air – a previously unknown source of human exposure to the biocide 2-mercaptobenzothiazole. *Environmental Science and Pollution Research International* 21: 11580–6.

Barnes, J.H., Chatterton, T.J. & Longhurst, J.W.S. (2019) Emissions vs exposure: increasing injustice from road traffic-related air pollution in the United Kingdom. *Transportation Research Part D: Transport and Environment* 73: 56–66.

Borroni, E., Pesatori, A.C., Bollati, V., Buoli, M. & Carugno, M. (2021) Air pollution exposure and depression: a comprehensive updated systematic review and meta-analysis. *Environmental Pollution*: 118245.

Brunekreef, B., Beelen, R., Hoek, G., Schouten, L., Bausch-Goldbohm, S., Fischer, P., Armstrong, B., Hughes, E., Jerrett, M. & van den Brandt, P. (2009) Effects of long-term exposure to traffic-related air pollution on respiratory and cardiovascular mortality in the Netherlands: the NLCS-AIR study. *Res Rep Health Eff Inst*: 5–71; discussion 73–89.

Burnett, R., Chen, H., Szyszkowicz, M., Fann, N., Hubbell, B., Pope, C.A., Apte, J.S., Brauer, M., Cohen, A., Weichenthal, S., *et al.* (2018) Global estimates of mortality associated with long-term exposure to outdoor fine particulate matter. *Proceedings of the National Academy of Sciences of the USA* 115: 9592.

Danopoulos, E., Twiddy, M., West, R. & Rotchell, J.M. (2021) A rapid review and meta-regression analyses of the toxicological impacts of microplastic exposure in human cells. *Journal of Hazardous Materials*: 127861.

Finkelstein, M.M., Jerrett, M. & Sears, M.R. (2004) Traffic air pollution and mortality rate advancement periods. *American Journal of Epidemiology* 160: 173–177.

Fuller, R., Landrigan, P.J., Balakrishnan, K., Bathan, G., Bose-O'Reilly, S., Brauer, M., Caravanos, J., Chiles, T., Cohen, A., Corra, L., *et al.* (2022) Pollution and health: a progress update. *The Lancet Planetary Health*, 10.1016/S2542-5196(22)00090-0.

Kheirbek, I., Haney, J., Douglas, S., Ito, K. & Matte, T. (2016) The contribution of motor vehicle emissions to ambient fine particulate matter public health impacts in New York City: a health burden assessment. *Environ Health* 15: 89.

Kwon, H.-S., Ryu, M.H. & Carlsten, C. (2020) Ultrafine particles: unique physicochemical properties relevant to health and disease. *Experimental & Molecular Medicine* 52: 318–328.

Loopmans, M., Smits, L. & Kenis, A. (2021) Rethinking environmental justice: capability building, public knowledge and the struggle against traffic-related air pollution. *Environment and Planning C: Politics and Space*: 23996544211042876.

Mak, K.L. & Loh, W.K.A. (2022) Relationship between road traffic exposure and human health. *European Journal of Environment and Earth Sciences* 3: 68–72.

Mudway, I.S., Dundas, I., Wood, H.E., Marlin, N., Jamaludin, J.B., Bremner, S.A., Cross, L., Grieve, A., Nanzer, A., Barratt, B.M., Beevers, S., Dajnak, D., Fuller, G.W., Font, A., Colligan, G., Sheikh, A., Walton, R., Grigg, J., Kelly, F.J., Lee, T.H. & Griffiths, C.J. (2019) Impact of London's low emission zone on air quality and children's respiratory health: a sequential annual cross-sectional study. *The Lancet Public Health* 4: e28-e40.

OECD (2014) *The Cost of Air Pollution: Health Impacts of Road Transport*. Paris: OECD Publishing.

Salmón, P., Stroh, E., Herrera-Dueñas, A., von Post, M. & Isaksson, C. (2018) Oxidative stress in birds along a NOx and urbanisation gradient: an interspecific approach. *Science of The Total Environment* 622–623: 635–643.

Sanderfoot, O.V. & Holloway, T. (2017) Air pollution impacts on avian species via inhalation exposure and associated outcomes. *Environmental Research Letters* 12: 083002.

Skakkebæk, N.E., Lindahl-Jacobsen, R., Levine, H., Andersson, A.-M., Jørgensen, N., Main, K.M., Lidegaard, Ø., Priskorn, L., Holmboe, S.A., Bräuner, E.V., Almstrup, K., Franca, L.R., Znaor, A., Kortenkamp, A., Hart, R.J. & Juul, A. (2021) Environmental factors in declining human fertility. *Nature Reviews Endocrinology* 18: 139–157.

Stenson, C., Wheeler, A.J., Carver, A., Donaire-Gonzalez, D., Alvarado-Molina, M., Nieuwenhuijsen, M. & Tham, R. (2021) The impact of traffic-related air pollution on child and adolescent academic performance: a systematic review. *Environment International* 155: 106696.

Wei, F., Yu, Z., Zhang, X., Wu, M., Wang, J., Shui, L., Lin, H., Jin, M., Tang, M. & Chen, K. (2021) Long-term exposure to ambient air pollution and incidence of depression: a population-based cohort study in China. *Science of The Total Environment*: 149986.

World Health Organisation Europe (2005) *Health Effects of Transport-Related Air Pollution*. Copenhagen: World Health Organisation.

Yim, S.H.L. & Barrett, S.R.H. (2012) Public health impacts of combustion emissions in the United Kingdom. *Environmental Science & Technology* 46: 4291–4296.

You, R., Ho, Y.-S. & Chang, R.C.-C. (2022) The pathogenic effects of particulate matter on neurodegeneration: a review. *Journal of Biomedical Science* 29: 15.

Zhang, K. & Batterman, S. (2013) Air pollution and health risks due to vehicle traffic. *Science of the Total Environment* 450–451: 307–316.

Zhong, N., Cao, J. & Wang, Y. (2017) Traffic congestion, ambient air pollution, and health: evidence from driving restrictions in Beijing. *Journal of the Association of Environmental and Resource Economists* 4: 821–856.

COVID severity and air quality

Bourdrel, T., Annesi-Maesano, I., Alahmad, B., Maesano, C.N. & Bind, M.-A. (2021) The impact of outdoor air pollution on COVID-19: a review of evidence from *in vitro* animal, and human studies. *European Respiratory Review* 30: 200242.

Conte, M.N., Gordon, M., Swartwood, N.A., Wilwerding, R. & Yu, C.A. (2021) The causal effects of chronic air pollution on the intensity of COVID-19 disease: some answers are blowing in the wind. *medRxiv*: 2021.04.28.21256146.

Konstantinoudis, G., Padellini, T., Bennett, J., Davies, B., Ezzati, M. & Blangiardo, M. (2021) Long-term exposure to air-pollution and COVID-19 mortality in England: a hierarchical spatial analysis. *Environment International* 146: 106316.

Perera, F., Berberian, A., Cooley, D., Shenaut, E., Olmstead, H., Ross, Z. & Matte, T. (2021) Potential health benefits of sustained air quality improvements in New York City: a simulation based on air pollution levels during the COVID-19 shutdown. *Environmental Research* 193: 110555.

Pozzer, A., Dominici, F., Haines, A., Witt, C., Münzel, T. & Lelieveld, J. (2020) Regional and global contributions of air pollution to risk of death from COVID-19. *Cardiovascular Research* 116: 2247–2253.

Sanchez-Piedra, C., Cruz-Cruz, C., Gamiño-Arroyo, A.-E. & Prado-Galbarro, F.-J. (2021) Effects of air pollution and climatology on COVID-19 mortality in Spain. *Air Quality, Atmosphere & Health* 14: 1869–1875.

Travaglio, M., Yu, Y., Popovic, R., Selley, L., Leal, N.S. & Martins, L.M. (2021) Links between air pollution and COVID-19 in England. *Environmental Pollution* 268: 115859.

Walton, H., Evangelopoulos, D., Kasdagli, M., Selley, L., Dajnak, D. & Katsouyanni, K. (2021) *Investigating Links between Air Pollution, COVID-19 and Lower Respiratory Infectious Diseases*. London: Imperial College Environmental Research Group.

Wu, X., Nethery, R.C., Sabath, M.B., Braun, D. & Dominici, F. (2020) Air pollution and COVID-19 mortality in the United States: strengths and limitations of an ecological regression analysis. *Science Advances* 6: eabd4049.

Xin, Y., Shao, S., Wang, Z., Xu, Z. & Li, H. (2021) COVID-2019 lockdown in Beijing: a rare opportunity to analyze the contribution rate of road traffic to air pollutants. *Sustainable Cities and Society* 75: 102989.

Yao, Y., Pan, J., Liu, Z., Meng, X., Wang, W., Kan, H. & Wang, W. (2021) Ambient nitrogen dioxide pollution and spreadability of COVID-19 in Chinese cities. *Ecotoxicology and Environmental Safety* 208: 111421.

Zhang, X., Tang, M., Guo, F., Wei, F., Yu, Z., Gao, K., Jin, M., Wang, J. & Chen, K. (2021) Associations between air pollution and COVID-19 epidemic during quarantine period in China. *Environmental Pollution* 268: 115897.

Impacts of traffic pollution on roadside wildlife

Bignal, K.L., Ashmore, M.R. & Power, S. (2004) *The Ecological Effects of Diffuse Air Pollution from Road Transport. Report no. 580.* Peterborough: English Nature.

Chameides, W.L., Yu, H., Liu, S.C., Bergin, M., Zhou, X., Mearns, L., Wang, G., Kiang, C.S., Saylor, R.D., Luo, C., Huang, Y., Steiner, A. & Giorgi, F. (1999) Case study of the effects of atmospheric aerosols and regional haze on agriculture: an opportunity to enhance crop yields in China through emission controls? *Proceedings of the National Academy of Sciences* 96: 13626–13633.

Frati, L., Caprasecca, E., Santoni, S., Gaggi, C., Guttova, A., Gaudino, S., Pati, A., Rosamilia, S., Pirintsos, S.A. & Loppi, S. (2006) Effects of NO_2 and NH_3 from road traffic on epiphytic lichens. *Environmental Pollution* 142: 58–64.

Girling, R.D., Lusebrink, I., Farthing, E., Newman, T.A. & Poppy, G.M. (2013) Diesel exhaust rapidly degrades floral odours used by honeybees. *Scientific Reports* 3: 2779.

Jürgens, A. & Bischoff, M. (2017) Changing odour landscapes: the effect of anthropogenic volatile pollutants on plant–pollinator olfactory communication. *Functional Ecology* 31: 56–64.

Leonard, R., Vergoz, V., Proschogo, N., McArthur, C. & Hochuli, D. (2018) Petrol exhaust pollution impairs honey bee learning and memory. *Oikos* 128: 264–273.

Leonard, R., Pettit, T., Irga, P., McArthur, C. & Hochuli, D. (2019) Acute exposure to urban air pollution impairs olfactory learning and memory in honeybees. *Ecotoxicology* 28: 1056–1062

Liang, Y., Rudik, I., Zou, E.Y., Johnston, A., Rodewald, A.D. & Kling, C.L. (2020) Conservation cobenefits from air pollution regulation: evidence from birds. *Proceedings of the National Academy of Sciences* 117: 30900.

Lobell, D.B., Di Tommaso, S. & Burney Jennifer, A. (2022) Globally ubiquitous negative effects of nitrogen dioxide on crop growth. *Science Advances* 8: eabm9909.

Muskett, C.J. & Jones, M.P. (1980) The dispersal of lead, cadmium and nickel from motor vehicles and effects on roadside invertebrate macrofauna. *Environmental Pollution Series A, Ecological and Biological* 23: 231–242.

Papa, G., Capitani, G., Capri, E., Pellecchia, M. & Negri, I. (2021) Vehicle-derived ultrafine particulate contaminating bees and bee products. *Science of The Total Environment* 750: 141700.

Peach, W.J., Vincent, K.E., Fowler, J.A. & Grice, P.V. (2008) Reproductive success of house sparrows along an urban gradient. *Animal Conservation* 11: 493–503.

Phillips, B.B., Bullock, J.M., Gaston, K.J., Hudson-Edwards, K.A., Bamford, M., Cruse, D., Dicks, L.V., Falagan, C., Wallace, C. & Osborne, J.L. (2021) Impacts of multiple pollutants on pollinator activity in road verges. *Journal of Applied Ecology* 58: 1017–1029.

Pratt, C. & Lottermoser, B. (2007) Trace metal uptake by the grass *Melinis repens* from roadside soils and sediments, tropical Australia. *Environmental Geology* 52: 1651–1662.

Ricardo-AEA (2016) *Potential Risk of Impacts of Nitrogen Oxides from Road Traffic on Designated Nature Conservation Sites*. Natural England Commissioned Report NECR200.

Ryalls, J.M.W., Langford, B., Mullinger, N.J., Bromfield, L.M., Nemitz, E., Pfrang, C. & Girling, R.D. (2022) Anthropogenic air pollutants reduce insect-mediated pollination services. *Environmental Pollution* 297: 118847.

Smithers, R., Harris, R. & Hitchcock, G. (2016) *The Ecological Effects of Air Pollution from Road Transport: an Updated Review*. UK: Natural England.

Spósito, J.C., Crispim Bdo, A., Mussury, R.M. & Grisolia, A.B. (2015) Genetic instability in plants associated with vehicular traffic and climatic variables. *Ecotoxicology and Environmental Safety* 120: 445–8.

Thimmegowda, G.G., Mullen, S., Sottilare, K., Sharma, A., Mohanta, S.S., Brockmann, A., Dhandapany, P.S. & Olsson, S.B. (2020) A field-based quantitative analysis of sublethal effects of air pollution on pollinators. *Proceedings of the National Academy of Sciences* 117: 20653–20661.

Tie, X., Huang, R.-J., Dai, W., Cao, J., Long, X., Su, X., Zhao, S., Wang, Q. & Li, G. (2016) Effect of heavy haze and aerosol pollution on rice and wheat productions in China. *Scientific Reports* 6: 29612.

Road salt pollution

Brady, S.P., Goedert, D., Frymus, L.E., Zamora-Camacho, F.J., Smith, P.C., Zeiss, C.J., Comas, M., Abbott, T.A., Basu, S.P., DeAndressi, J.C., Forgione, M.E., Maloney, M.J., Priester, J.L., Senturk, F., Szeligowski, R.V., Tucker, A.S., Zhang, M. & Calsbeek, R. (2022) Salted roads lead to oedema and reduced locomotor function in amphibian populations. *Freshwater Biology* 67: 1150–1161.

Braun, S. & Flückiger, W. (1984) Increased population of the aphid *Aphis pomi* at a motorway. Part 2—The effect of drought and deicing salt. *Environmental Pollution Series A, Ecological and Biological* 36: 261–270.

Dananay, K.L., Krynak, K.L., Krynak, T.J. & Benard, M.F. (2015) Legacy of road salt: apparent positive larval effects counteracted by negative postmetamorphic effects in wood frogs. *Environmental Toxicology and Chemistry* 34: 2417–2424.

Denoël, M., Bichot, M., Ficetola, G.F., Delcourt, J., Ylieff, M., Kestemont, P. & Poncin, P. (2010) Cumulative effects of road de-icing salt on amphibian behavior. *Aquatic Toxicology* 99: 275–280.

Dugan, H.A., Bartlett, S.L., Burke, S.M., Doubek, J.P., Krivak-Tetley, F.E., Skaff, N.K., Summers, J.C., Farrell, K.J., McCullough, I.M., Morales-Williams, A.M., Roberts, D.C., Ouyang, Z., Scordo, F., Hanson, P.C. & Weathers, K.C. (2017) Salting our freshwater lakes. *Proceedings of the National Academy of Sciences* 114: 4453.

Fay, L. & Shi, X. (2012) Environmental impacts of chemicals for snow and ice control: state of the knowledge. *Water Air and Soil Pollution* 223: 2751–2770.

Findlay, S.E. & Kelly, V.R. (2011) Emerging indirect and long-term road salt effects on ecosystems. *Annals of the New York Academy of Science* 1223: 58–68.

Frymus, L.E., Goedert, D., Zamora-Camacho, F.J., Smith, P.C., Zeiss, C.J., Comas, M., Abbott, T.A., Basu, S.P., DeAndressi, J.C., Forgione, M.E., Maloney, M.J., Priester, J.L., Senturk, F., Szeligowski, R.V., Tucker, A.S., Zhang, M., Calsbeek, R. & Brady, S.P. (2021) Salted roads lead to edema and reduced locomotor function in wood frogs. *bioRxiv*: 2021.03.23.436008.

Grosman, P.D., Jaeger, J.A.G., Pascale, B., Dussault, C. & Ouellet, J.P. (2009) Reducing moose-vehicle collisions through salt pool removal and displacement: an agent-based modeling approach. *Ecology and Society* 14: 17.

Hall, E.M., Brunner, J.L., Hutzenbiler, B. & Crespi, E.J. (2020) Salinity stress increases the severity of ranavirus epidemics in amphibian populations. *Proceedings of the Royal Society B: Biological Sciences* 287: 20200062.

Hintz, W.D. & Relyea, R.A. (2019) A review of the species, community, and ecosystem impacts of road salt salinisation in fresh waters. *Freshwater Biology* 64: 1081–1097.

Hintz, W.D., Fay, L. & Relyea, R.A. (2021) Road salts, human safety, and the rising salinity of our fresh waters. *Frontiers in Ecology and the Environment* 20: 22–30.

Karraker, N.E. & Ruthig, G.R. (2009) Effect of road deicing salt on the susceptibility of amphibian embryos to infection by water molds. *Environmental Research* 109: 40–45.

Karraker, N.E. & Gibbs, J.P. (2011) Road deicing salt irreversibly disrupts osmoregulation of salamander egg clutches. *Environmental Pollution* 159: 833–835.

Keerti, K., Jackson, S., Taylor, G., Hannah, T. & Víctor, D.C.-G. (2021) A preliminary evaluation of the environmental toxicology of road salts on plant root dynamics. *BIOS* 92: 48–51.

Lambert, M.R., Stoler, A.B., Smylie, M.S., Relyea, R.A. & Skelly, D.K. (2016) Interactive effects of road salt and leaf litter on wood frog sex ratios and sexual size dimorphism. *Canadian Journal of Fisheries and Aquatic Sciences* 74: 141–146.

Lawson, L. & Jackson, D.A. (2021) Salty summertime streams—road salt contaminated watersheds and estimates of the proportion of impacted species. *FACETS* 6: 317–333.

Leggett, S., Borrelli, J., Jones, D.K. & Relyea, R. (2021) The combined effects of road salt and biotic stressors on amphibian sex ratios. *Environmental Toxicology and Chemistry* 40: 231–235.

Mineau, P. & Brownlee, L.J. (2005) Road salts and birds: an assessment of the risk with particular emphasis on winter finch mortality. *Wildlife Society Bulletin* 33: 835–841.

Petranka, J.W. & Doyle, E.J. (2010) Effects of road salts on the composition of seasonal pond communities: can the use of road salts enhance mosquito recruitment? *Aquatic Ecology* 44: 155–166.

Schuler, M.S., Hintz, W.D., Jones, D.K., Lind, L.A., Mattes, B.M., Stoler, A.B., Sudol, K.A. & Relyea, R.A. (2017) How common road salts and organic additives alter freshwater food webs: in search of safer alternatives. *Journal of Applied Ecology* 54: 1353–1361.

Sibert, R., Koretsky, C. & Wyman, D. (2014) Cultural meromixis: effects of road-salt on the chemical stratification of an urban kettle lake. *Chemical Geology* 395.

Stockwell, M.P., Clulow, J. & Mahony, M.J. (2015) Evidence of a salt refuge: chytrid infection loads are suppressed in hosts exposed to salt. *Oecologia* 177: 901–910.

Töpfer, T., Mazánek, L. & Bureš, S. (2014) Deadly gastroliths: Eurasian siskins *Carduelis spinus* poisoned by road salt grains. *Ardea* 102: 101–104.

Roads and light pollution

Aulsebrook, A.E., Connelly, F., Johnsson, R.D., Jones, T.M., Mulder, R.A., Hall, M.L., Vyssotski, A.L. & Lesku, J.A. (2020) White and amber light at night disrupt sleep physiology in birds. *Current Biology* 30: 3657–3663.e5.

Aulsebrook, A.E., Johnsson, R.D. & Lesku, J.A. (2021) Light, sleep and performance in diurnal birds. *Clocks & Sleep* 3.

Bennie, J., Duffy, J.P., Davies, T.W., Correa-Cano, M.E. & Gaston, K.J. (2015) Global trends in exposure to light pollution in natural terrestrial ecosystems. *Remote Sensing* 7: 10.3390/rs70302715.

Bennie, J., Davies, T.W., Cruse, D. & Gaston, K.J. (2016) Ecological effects of artificial light at night on wild plants. *Journal of Ecology* 104: 611–620.

Bennie, J., Davies, T.W., Cruse, D., Inger, R. & Gaston, K.J. (2018) Artificial light at night causes top-down and bottom-up trophic effects on invertebrate populations. *Journal of Applied Ecology* 55: 2698–2706.

Blume, C., Garbazza, C. & Spitschan, M. (2019) Effects of light on human circadian rhythms, sleep and mood. *Somnologie (Berl)* 23: 147–156.

Boyes, D.H., Evans, D.M., Fox, R., Parsons, M.S. & Pocock, M.J.O. (2021a) Street lighting has detrimental impacts on local insect populations. *Science Advances*: DOI: 10.1126/sciadv.abi8322.

Boyes, D.H., Evans, D.M., Fox, R., Parsons, M.S. & Pocock, M.J.O. (2021b) Is light pollution driving moth population declines? A review of causal mechanisms across the life cycle. *Insect Conservation and Diversity* 14: 167–187.

Cox, D.T.C., Sánchez de Miguel, A., Dzurjak, S.A., Bennie, J. & Gaston, K.J. (2020) National scale spatial variation in artificial light at night. *Remote Sensing,* 12, 10.3390/rs12101591.

Da Silva, A., Samplonius, J.M., Schlicht, E., Valcu, M. & Kempenaers, B. (2014) Artificial night lighting rather than traffic noise affects the daily timing of dawn and dusk singing in common European songbirds. *Behavioral Ecology* 25: 1037–1047.

Da Silva, A., Valcu, M. & Kempenaers, B. (2015) Light pollution alters the phenology of dawn and dusk singing in common European songbirds. *Philosophical Transactions of the Royal Society B: Biological Sciences* 370: 20140126.

Davies, T.W. & Smyth, T. (2018) Why artificial light at night should be a focus for global change research in the 21st century. *Global Change Biology* 24: 872–882.

Deichmann, J.L., Ampudia Gatty, C., Andía Navarro, J.M., Alonso, A., Linares-Palomino, R. & Longcore, T. (2021) Reducing the blue spectrum of artificial light at night minimises insect attraction in a tropical lowland forest. *Insect Conservation and Diversity* 14: 247–259.

Ditmer, M.A., Stoner, D.C. & Carter, N.H. (2021a) Estimating the loss and fragmentation of dark environments in mammal ranges from light pollution. *Biological Conservation* 257: 109135.

Ditmer, M.A., Stoner, D.C., Francis, C.D., Barber, J.R., Forester, J.D., Choate, D.M., Ironside, K.E., Longshore, K.M., Hersey, K.R., Larsen, R.T., McMillan, B.R., Olson, D.D., Andreasen, A.M., Beckmann, J.P., Holton, P.B., Messmer, T.A. & Carter, N.H. (2021b) Artificial nightlight alters the predator–prey dynamics of an apex carnivore. *Ecography* 44: 149–161.

Dwyer, R.G., Bearhop, S., Campbell, H.A. & Bryant, D.M. (2013) Shedding light on light: benefits of anthropogenic illumination to a nocturnally foraging shorebird. *Journal of Animal Ecology* 82: 478–85.

Falchi, F., Cinzano, P., Duriscoe, D., Kyba, C.C.M., Elvidge, C.D., Baugh, K., Portnov, B.A., Rybnikova, N.A. & Furgoni, R. (2016) The new world atlas of artificial night sky brightness. *Science Advances* 2: e1600377.

ffrench-Constant, R.H., Somers-Yeates, R., Bennie, J., Economou, T., Hodgson, D., Spalding, A. & McGregor, P.K. (2016) Light pollution is associated with earlier tree budburst across the United Kingdom. *Proceedings of the Royal Society B: Biological Sciences* 283: 20160813.

Foster, J.J., Tocco, C., Smolka, J., Khaldy, L., Baird, E., Byrne, M.J., Nilsson, D.-E. & Dacke, M. (2021) Light pollution forces a change in dung beetle orientation behavior. *Current Biology* 31: 3935–3942.

Garrett, J.K., Donald, P.F. & Gaston, K.J. (2020) Skyglow extends into the world's Key Biodiversity Areas. *Animal Conservation* 23: 153–159.

Gaston, K.J., Bennie, J., Davies, T.W. & Hopkins, J. (2013) The ecological impacts of nighttime light pollution: a mechanistic appraisal. *Biological Reviews* 88: 912–927.

Gaston, K.J., Davies, T.W., Nedelec, S.L. & Holt, L.A. (2017) Impacts of artificial light at night on biological timings. *Annual Review of Ecology, Evolution, and Systematics* 48: 49–68.

Gaston, K.J. & Holt, L.A. (2018) Nature, extent and ecological implications of night-time light from road vehicles. *Journal of Applied Ecology* 55: 2296–2307.

Gaston, K.J., Ackermann, S., Bennie, J., Cox, D.T.C., Phillips, B.B., Sánchez de Miguel, A. & Sanders, D. (2021) Pervasiveness of biological impacts of artificial light at night. *Integrative and Comparative Biology* 61: 1098–1110.

Grubisic, M. & van Grunsven, R.H.A. (2021) Artificial light at night disrupts species interactions and changes insect communities. *Current Opinion in Insect Science* 47: 136–141.

Hölker, F., Wolter, C., Perkin, E.K. & Tockner, K. (2010) Light pollution as a biodiversity threat. *Trends in Ecology and Evolution* 25: 681–682.

Horton, K.G., Nilsson, C., Van Doren, B.M., La Sorte, F.A., Dokter, A.M. & Farnsworth, A. (2019) Bright lights in the big cities: migratory birds' exposure to artificial light. *Frontiers in Ecology and the Environment* 17: 209–214.

Injaian, A.S., Uehling, J.J., Taff, C.C. & Vitousek, M.N. (2021) Effects of artificial light at night on avian provisioning, corticosterone, and reproductive success. *Integrative and Comparative Biology*, 10.1093/icb/icab055.

Kempenaers, B., Borgström, P., Loës, P., Schlicht, E. & Valcu, M. (2010) Artificial night lighting affects dawn song, extra-pair siring success, and lay date in songbirds. *Current Biology* 20: 1735–1739.

Knop, E., Zoller, L., Ryser, R., Gerpe, C., Hörler, M. & Fontaine, C. (2017) Artificial light at night as a new threat to pollination. *Nature* 548: 206–209.

Kyba, C.C.M. & Hölker, F. (2013) Do artificially illuminated skies affect biodiversity in nocturnal landscapes? *Landscape Ecology* 28: 1637–1640.

Lewanzik, D. & Voigt, C.C. (2014) Artificial light puts ecosystem services of frugivorous bats at risk. *Journal of Applied Ecology* 51: 388–394.

Longcore, T. (2010) Sensory ecology: night lights alter reproductive behavior of blue tits. *Current Biology* 20: R893-R895.

Macgregor, C.J., Pocock, M.J.O., Fox, R. & Evans, D.M. (2019) Effects of street lighting technologies on the success and quality of pollination in a nocturnally pollinated plant. *Ecosphere* 10: e02550.

McLaren, J.D., Buler, J.J., Schreckengost, T., Smolinsky, J.A., Boone, M., Emiel van Loon, E., Dawson, D.K. & Walters, E.L. (2018) Artificial light at night confounds broad-scale habitat use by migrating birds. *Ecology Letters* 21: 356–364.

Ouyang, J.Q., de Jong, M., van Grunsven, R.H.A., Matson, K.D., Haussmann, M.F., Meerlo, P., Visser, M.E. & Spoelstra, K. (2017) Restless roosts: Light pollution affects behavior, sleep, and physiology in a free-living songbird. *Global Change Biology* 23: 4987–4994.

Owens, A.C.S., Cochard, P., Durrant, J., Farnworth, B., Perkin, E.K. & Seymoure, B. (2020) Light pollution is a driver of insect declines. *Biological Conservation* 241: 108259.

Rivera, G. & Borchert, R. (2001) Induction of flowering in tropical trees by a 30-min reduction in photoperiod: evidence from field observations and herbarium specimens. *Tree Physiology* 21: 201–212.

Rodríguez, A. & Rodríguez, B. (2009) Attraction of petrels to artificial lights in the Canary Islands: effects of the moon phase and age class. *Ibis* 151: 299–310.

Salinas-Ramos, V.B., Ancillotto, L., Cistrone, L., Nastasi, C., Bosso, L., Smeraldo, S., Sánchez Cordero, V. & Russo, D. (2021) Artificial illumination influences niche segregation in bats. *Environmental Pollution* 284: 117187.

Sanders, D., Kehoe, R., Tiley, K., Bennie, J., Cruse, D., Davies, T.W., Frank van Veen, F.J. & Gaston, K.J. (2015) Artificial nighttime light changes aphid-parasitoid population dynamics. *Scientific Reports* 5: 15232.

Sanders, D., Frago, E., Kehoe, R., Patterson, C. & Gaston, K.J. (2021) A meta-analysis of biological impacts of artificial light at night. *Nature Ecology & Evolution* 5: 74–81.

Silva, E., Marco, A., da Graça, J., Pérez, H., Abella, E., Patino-Martinez, J., Martins, S. & Almeida, C. (2017) Light pollution affects nesting behavior of loggerhead turtles and predation risk of nests and hatchlings. *Journal of Photochemistry and Photobiology B: Biology* 173: 240–249.

Stewart, A.J.A. (2021) Impacts of artificial lighting at night on insect conservation. *Insect Conservation and Diversity* 14: 163–166.

Stone, E.L., Jones, G. & Harris, S. (2012) Conserving energy at a cost to biodiversity? Impacts of LED lighting on bats. *Global Change Biology* 18: 2458–2465.

Van den Broeck, M., De Cock, R., Van Dongen, S. & Matthysen, E. (2021) Blinded by the light: artificial light lowers mate attraction success in female glow-worms (*Lampyris noctiluca* L.). *Insects* 12: 734.

van Geffen, K.G., van Eck, E., de Boer, R.A., van Grunsven, R.H.A., Salis, L., Berendse, F. & Veenendaal, E.M. (2015) Artificial light at night inhibits mating in a geometrid moth. *Insect Conservation and Diversity* 8: 282–287.

van Langevelde, F., Ettema, J.A., Donners, M., WallisDeVries, M.F. & Groenendijk, D. (2011) Effect of spectral composition of artificial light on the attraction of moths. *Biological Conservation* 144: 2274–2281.

Willems, J.S., Phillips, J.N. & Francis, C.D. (2022) Artificial light at night and anthropogenic noise alter the foraging activity and structure of vertebrate communities. *Science of the Total Environment* 805: 150223.

Wilson, A.A., Ditmer, M.A., Barber, J.R., Carter, N.H., Miller, E.T., Tyrrell, L.P. & Francis, C.D. (2021) Artificial night light and anthropogenic noise interact to influence bird abundance over a continental scale. *Global Change Biology* 27: 3987–4004.

Impacts of light pollution on human health

Cho, Y., Ryu, S.H., Lee, B.R., Kim, K.H., Lee, E. & Choi, J. (2015) Effects of artificial light at night on human health: A literature review of observational and experimental studies applied to exposure assessment. *Chronobiology International* 32: 1294–310.

He, S., Shao, W. & Han, J. (2021) Have artificial lighting and noise pollution caused zoonosis and the COVID-19 pandemic? A review. *Environmental Chemistry Letters*.

Münzel, T., Hahad, O. & Daiber, A. (2021) The dark side of nocturnal light pollution. Outdoor light at night increases risk of coronary heart disease. *European Heart Journal* 42: 831–834.

Chapter 8: In the Zone

Andrasi, B., Jaeger, J.A.G., Heinicke, S., Metcalfe, K. & Hockings, K.J. (2021) Quantifying the road-effect zone for a critically endangered primate. *Conservation Letters* 2021: e12839.

Arévalo, J.E. & Newhard, K. (2011) Traffic noise affects forest bird species in a protected tropical forest. *Revista de Biología Tropical* 59: 969–980.

Berthinussen, A. & Altringham, J. (2012) The effect of a major road on bat activity and diversity. *Journal of Applied Ecology* 49: 82–89.

Bhardwaj, M., Soanes, K., Lahoz-Monfort, J.J., Lumsden, L.F. & van der Ree, R. (2021) Insectivorous bats are less active near freeways. *PLOS ONE* 16: e0247400.

Bissonette, J.A. & Rosa, S.A. (2009) Road zone effects in small-mammal communities. *Ecology and Society* 14: 10.5751/ES-02753-140127.

Boarman, W.I. & Sazaki, M. (2006) A highway's road-effect zone for desert tortoises (*Gopherus agassizii*). *Journal of Arid Environments* 65: 94–101.

Brotons, L. & Herrando, S. (2001) Reduced bird occurrence in pine forest fragments associated with road proximity in a Mediterranean agricultural area. *Landscape and Urban Planning* 57: 77–89.

Cardoso, G.C., Klingbeil, B.T., La Sorte, F.A., Lepczyk, C.A., Fink, D. & Flather, C.H. (2020) Exposure to noise pollution across North American passerines supports the noise filter hypothesis. *Global Ecology and Biogeography* 29: 1430–1434.

Chen, H.L. & Koprowski, J.L. (2015) Animal occurrence and space use change in the landscape of anthropogenic noise. *Biologcal Conservation* 192: 315–322.

D'Amico, M., Périquet, S., Román, J. & Revilla, E. (2016) Road avoidance responses determine the impact of heterogeneous road networks at a regional scale. *Journal of Applied Ecology* 53: 181–190.

Eigenbrod, F., Hecnar, S.J. & Fahrig, L. (2008) The relative effects of road traffic and forest cover on anuran populations. *Biological Conservation* 141: 35–46.

Eigenbrod, F., Hecnar, S.J. & Fahrig, L. (2009) Quantifying the road-effect zone: threshold effects of a motorway on anuran populations in Ontario, Canada. *Ecology and Society* 14: 24.

Fahrig, L., Pedlar, J.H., Pope, S.E., Taylor, P.D. & Wegner, J.F. (1995) Effect of road traffic on amphibian density. *Biological Conservation* 73: 177–182.

Forman, R.T.T. (2000) Estimate of the area affected ecologically by the road system in the United States. *Conservation Biology* 14: 31–35.

Forman, R.T.T. & Deblinger, R.D. (2000) The ecological road-effect zone of a Massachusetts (USA) suburban highway. *Conservation Biology* 14: 36–46.

Forman, R.T.T., Reineking, B. & Hersperger, A.M. (2002) Road traffic and nearby grassland bird patterns in a suburbanizing landscape. *Environmental Management* 29: 782–800.

Francis, C.D. (2015) Vocal traits and diet explain avian sensitivities to anthropogenic noise. *Global Change Biology* 21: 1809–1820.

Goodwin, S.E. & Shriver, W.G. (2011) Effects of traffic noise on occupancy patterns of forest birds. *Conservation Biology* 25: 406–411.

Griffith, E.H., Sauer, J.R. & Royle, J.A. (2010) Traffic effects on bird counts on North American Breeding Bird Survey routes. *Auk* 127: 387–393.

Haskell, D.G. (2000) Effects of forest roads on macroinvertebrate soil fauna of the southern Appalachian mountains. *Conservation Biology* 14: 57–63.

Helldin, J.O., Collinder, P., Bengtsson, D., Karlberg, A. & Askling, J. (2013) Assessment of traffic noise impact in important bird sites in Sweden – a practical method for the regional scale. *Oecologia Australis* 17: 48–62.

Helldin, J.O. (2019) Predicted impacts of transport infrastructure and traffic on bird conservation in Swedish Special Protection Areas. *Nature Conservation* 36: 1–16.

Husby, M. (2017) Traffic influence on roadside bird abundance and behaviour. *Acta Ornithologica* 52: 93–103.

Iglesias-Merchán, C., Diaz-Balteiro, L. & de la Puente, J. (2016) Road traffic noise impact assessment in a breeding colony of cinereous vultures (*Aegypius monachus*) in Spain. *The Journal of the Acoustical Society of America* 139: 1124–1131.

Jack, J., Rytwinski, T., Fahrig, L. & Francis, C.M. (2015) Influence of traffic mortality on forest bird abundance. *Biodiversity and Conservation* 24: 1507–1529.

Kuitunen, M., Rossi, E. & Stenroos, A. (1998) Do highways influence density of land birds? *Environmental Management* 22: 297–302.

Leblond, M., Dussault, C. & Ouellet, J.P. (2013) Avoidance of roads by large herbivores and its relation to disturbance intensity. *Journal of Zoology* 289: 32–40.

Mammides, C., Kadis, C. & Coulson, T. (2015) The effects of road networks and habitat heterogeneity on the species richness of birds in Natura 2000 sites in Cyprus. *Landscape Ecology* 30: 67–75.

Mayer, M., Fischer, C., Blaum, N., Sunde, P. & Ullmann, W. (2022) Influence of roads on space use by European hares in different landscapes. *Research Square*: 10.21203/rs.3.rs-1501391/v1.

Milsom, T.P., Langton, S.D., Parkin, W.K., Allen, D.S., Bishop, J.D. & Hart, J.D. (2001) Coastal grazing marshes as a breeding habitat for skylarks *alauda arvensis*. In: *The Ecology and Conservation of Skylarks Alauda Arvensis,* ed. P.F. Donald & J.A. Vickery: pp. 41–51. Sandy: RSPB.

Nega, T., Smith, C., Bethune, J. & Fu, W.H. (2012) An analysis of landscape penetration by road infrastructure and traffic noise. *Computers Environment and Urban Systems* 36: 245–256.

Palomino, D. & Carrascal, L.M. (2007) Threshold distances to nearby cities and roads influence the bird community of a mosaic landscape. *Biological Conservation* 140: 100–109.

Peaden, J.M., Tuberville, T.D., Buhlmann, K.A., Nafus, M.G. & Todd, B.D. (2016) Delimiting road-effect zones for threatened species: implications for mitigation fencing. *Wildlife Research* 42: 650–659.

Peris, S.J. & Pescador, M. (2004) Effects of traffic noise on passerine populations in Mediterranean wooded pastures. *Applied Acoustics* 65: 357–366.

Polak, M., Wiacek, J., Kucharczyk, M. & Orzechowski, R. (2013) The effect of road traffic on a breeding community of woodland birds. *European Journal of Forest Research* 132: 931–941.

Proppe, D.S., Sturdy, C.B. & St. Clair, C.C. (2013) Anthropogenic noise decreases urban songbird diversity and may contribute to homogenization. *Global Change Biology* 19: 1075–1084.

Reijnen, R., Foppen, R., Terbraak, C. & Thissen, J. (1995) The effects of car traffic on breeding bird populations in woodland. 3. Reduction of density in relation to the proximity of main roads. *Journal of Applied Ecology* 32: 187–202.

Reijnen, R., Foppen, R. & Meeuwsen, H. (1996) The effects of traffic on the density of breeding birds in Dutch agricultural grasslands. *Biological Conservation* 75: 255–260.

Rheindt, F.E. (2003) The impact of roads on birds: Does song frequency play a role in determining susceptibility to noise pollution? *Journal Fur Ornithologie* 144: 295–306.

Roedenbeck, I.A. & Voser, P. (2008) Effects of roads on spatial distribution, abundance and mortality of brown hare (*Lepus europaeus*) in Switzerland. *European Journal of Wildlife Research* 54: 425–437.

Rytwinski, T. & Fahrig, L. (2011) Reproductive rate and body size predict road impacts on mammal abundance. *Ecological Applications* 21: 589–600.

Rytwinski, T. & Fahrig, L. (2012) Do species life history traits explain population responses to roads? A meta-analysis. *Biological Conservation* 147: 87–98.

Senzaki, M., Barber, J.R., Phillips, J.N., Carter, N.H., Cooper, C.B., Ditmer, M.A., Fristrup, K.M., McClure, C.J.W., Mennitt, D.J., Tyrrell, L.P., Vukomanovic, J., Wilson, A.A. & Francis, C.D. (2020a) Sensory pollutants alter bird phenology and fitness across a continent. *Nature* 587: 605–609.

Senzaki, M., Kadoya, T. & Francis, C.D. (2020b) Direct and indirect effects of noise pollution alter biological communities in and near noise-exposed environments. *Proceedings of the Royal Society B: Biological Sciences* 287: 20200176.

Shilling, F. & Waetjen, D. (2012) *The Road Effect Zone GIS Model*. Davis, CA: UC Davis Department of Environmental Science and Policy.

Silva, C.C., Lourenco, R., Godinho, S., Gomes, E., Sabino-Marques, H., Medinas, D., Neves, V., Silva, C., Rabaca, J.E. & Mira, A. (2012) Major roads have a negative impact on the tawny owl *Strix aluco* and the little owl *Athene noctua* populations. *Acta Ornithologica* 47: 47–54.

Speziale, K.L., Lambertucci, S.A. & Olsson, O. (2008) Disturbance from roads negatively affects Andean condor habitat use. *Biological Conservation* 141: 1765–1772.

Ste Marie, E., Turney, S. & Buddle, C. (2018) The effect of road proximity on arthropod communities in Yukon, Canada. *ARCTIC* 71: 89.

Summers, P.D., Cunnington, G.M. & Fahrig, L. (2011) Are the negative effects of roads on breeding birds caused by traffic noise? *Journal of Applied Ecology* 48: 1527–1534.

Tanner, D. & Perry, J. (2007) Road effects on abundance and fitness of Galapagos lava lizards (*Microlophus albemarlensis*). *Journal of Environmental Management* 85: 270–278.

Torres, A., Palacín, C., Seoane, J. & Alonso, J.C. (2011) Assessing the effects of a highway on a threatened species using before–during–after and before–during–after-control–impact designs. *Biological Conservation* 144: 2223–2232.

Torres, A., Jaeger, J.A.G. & Alonso, J.C. (2016) Assessing large-scale wildlife responses to human infrastructure development. *Proceedings of the National Academy of Sciences* 113: 8472.

van der Horst, S., Goytre, F., Marques, A., Santos, S., Mira, A. & Lourenço, R. (2019) Road effects on species abundance and population trend: a case study on tawny owl. *European Journal of Wildlife Research* 65: 99.

van der Vliet, R., Dijk, J. & Wassen, M. (2010) How different landscape elements limit the breeding habitat of meadow bird species. *Ardea* 98: 203–209.

van der Zande, A.N., ter Keurs, W.J. & van der Weijden, W.J. (1980) The impact of roads on the densities of four bird species in an open field habitat—evidence of a long-distance effect. *Biological Conservation* 18: 299–321.

Veen, J. (1973) The disturbance of meadow bird populations. *Stedebouw & Volkshuisvesting* 54: 16–26.

Watkins, R.Z., Chen, J., Pickens, J. & Brosofske, K.D. (2003) Effects of forest roads on understory plants in a managed hardwood landscape. *Conservation Biology* 17: 411–419.

Weiserbs, A. & Jacob, J.-P. (2001) Is breeding bird distribution affected by motorway traffic noise? *Alauda* 69: 483–489.

Whittington, J., Low, P. & Hunt, B. (2019) Temporal road closures improve habitat quality for wildlife. *Scientific Reports* 9: 3772.

Whitworth, A., Beirne, C., Rowe, J., Ross, F., Acton, C., Burdekin, O. & Brown, P. (2015) The response of faunal biodiversity to an unmarked road in the Western Amazon. *Biodiversity and Conservation* 24: 1657–1670.

Wilson, A.A., Ditmer, M.A., Barber, J.R., Carter, N.H., Miller, E.T., Tyrrell, L.P. & Francis, C.D. (2021) Artificial night light and anthropogenic noise interact to influence bird abundance over a continental scale. *Global Change Biology* 27: 3987–4004.

Chapter 9: The Sixth Horseman

Forman, R.T.T. (1998) Road ecology: a solution for the giant embracing us. *Landscape Ecology* 13: III–V.

Chapter 10: Winners and Losers

Ascensão, F., Clevenger, A.P., Grilo, C., Filipe, J. & Santos-Reis, M. (2012) Highway verges as habitat providers for small mammals in agrosilvopastoral environments. *Biodiversity and Conservation* 21: 3681–3697.

Blake, D., Hutson, A.M., Racey, P.A., Rydell, J. & Speakman, J.R. (1994) Use of lamplit roads by foraging bats in southern England. *Journal of Zoology* 234: 453–462.

Brischoux, F., Meillère, A., Dupoué, A., Lourdais, O. & Angelier, F. (2017) Traffic noise decreases nestlings' metabolic rates in an urban exploiter. *Journal of Avian Biology* 48: 905–909.

Burns, F., Eaton, M., Balmer, D., Banks, A., Caldow, R., Donelan, J., Douse, A., Duigan, C., Foster, S., Frost, T., Grice, P., Hall, C., Hanmer, H., Harris, S., Johnstone, I.,

Lindley, P., McCulloch, N., Noble, D., Risely, K., Robinson, R. & Wotton, S. (2020) *The state of the UK's birds 2020.* Sandy: The RSPB, BTO, WWT, DAERA, JNCC, NatureScot, NE and NRW.

Clark, W.D. & Karr, J.R. (1979) Effects of highways on red-winged blackbird and horned lark populations. *Wilson Bulletin* 91: 143–145.

Cooke, S.C., Balmford, A., Donald, P.F., Newson, S.E. & Johnston, A. (2020a) Roads as a contributor to landscape-scale variation in bird communities. *Nature Communications* 11: 3125.

Cooke, S.C., Balmford, A., Johnston, A., Newson, S.E. & Donald, P.F. (2020b) Variation in abundances of common bird species associated with roads. *Journal of Applied Ecology* 57: 1271–1282.

Crino, O.L., Johnson, E.E., Blickley, J.L., Patricelli, G.L. & Breuner, C.W. (2013) Effects of experimentally elevated traffic noise on nestling white-crowned sparrow stress physiology, immune function and life history. *Journal of Experimental Biology* 216: 2055–2062.

Dean, W.R.J. & Milton, S.J. (2003) The importance of roads and road verges for raptors and crows in the Succulent and Nama-Karoo, South Africa. *Ostrich* 74: 181–186.

Downing, R.J., Rytwinski, T. & Fahrig, L. (2015) Positive effects of roads on small mammals: a test of the predation release hypothesis. *Ecological Research* 30: 651–662.

Fekete, R., Nagy, T., Bódis, J., Biró, É., Löki, V., Süveges, K., Takács, A., Tökölyi, J. & Molnár V, A. (2017) Roadside verges as habitats for endangered lizard-orchids (*Himantoglossum* spp.): Ecological traps or refuges? *Science of The Total Environment* 607–608: 1001–1008.

Fekete, R., Haszonits, G., Schmidt, D., Bak, H., Vincze, O., Süveges, K. & Molnár V, A. (2021) Rapid continental spread of a salt-tolerant plant along the European road network. *Biological Invasions* 23: 2661–2674.

Fielding, M.W., Buettel, J.C., Brook, B.W., Stojanovic, D. & Yates, L.A. (2021) Roadkill islands: carnivore extinction shifts seasonal use of roadside carrion by generalist avian scavenger. *bioRxiv*: 2021.02.18.429855.

Friebe, K., Steffens, T., Schulz, B., Valqui, J., Reck, H. & Hartl, G. (2018) The significance of major roads as barriers and their roadside habitats as potential corridors for hazel dormouse migration – a population genetic study. *Folia Zoologica* 67: 98–109.

Grunst, M.L., Grunst, A.S., Pinxten, R. & Eens, M. (2021) Little parental response to anthropogenic noise in an urban songbird, but evidence for individual differences in sensitivity. *Science of The Total Environment* 769: 144554.

Halfwerk, W., Both, C. & Slabbekoorn, H. (2016) Noise affects nest-box choice of 2 competing songbird species, but not their reproduction. *Behavioral Ecology* 27: 1592–1600.

Hetherington, M., Sterling, P. & Coulthard, E. (2021) Butterfly colonisation of a new chalkland road cutting. *Insect Conservation and Diversity* 15: 191–199.

Hill, J.E., DeVault, T.L. & Belant, J.L. (2020) A review of ecological factors promoting road use by mammals. *Mammal Review* 51: 214–227.

Junker-Bornholdt, R., Wagner, M., Zimmermann, M., Simonis, S., Schmidt, K.-H. & Wiltschko, W. (1998) Zum Einfluß einer Autobahn im Bau und während des Betriebs auf die Brutbiologie von Kohlmeisen (*Parus major*) und Blaumeisen (*P. caeruleus*). *Journal für Ornithologie* 139: 131–139.

Kroeger, S.B., Hanslin, H.M., Lennartsson, T., D'Amico, M., Kollmann, J., Fischer, C., Albertsen, E. & Speed, J.D.M. (2021) Impacts of roads on bird species richness: a meta-analysis considering road types, habitats and feeding guilds. *Science of The Total Environment*: 151478.

Lambertucci, S.A., Speziale, K.L., Rogers, T.E. & Morales, J.M. (2009) How do roads affect the habitat use of an assemblage of scavenging raptors? *Biodiversity and Conservation* 18: 2063–2074.

Meunier, F.D., Verheyden, C. & Jouventin, P. (2000) Use of roadsides by diurnal raptors in agricultural landscapes. *Biological Conservation* 92: 291–298.

Morelli, F., Beim, M., Jerzak, L., Jones, D. & Tryjanowski, P. (2014) Can roads, railways and related structures have positive effects on birds? – A review. *Transportation Research Part D: Transport and Environment* 30: 21–31.

Murphy, R.E., Martin, A.E. & Fahrig, L. (2022) Reduced predation on roadside nests can compensate for road mortality in road-adjacent turtle populations. *Ecosphere* 13: e3946.

Phillips, B.B., Gaston, K.J., Bullock, J.M. & Osborne, J.L. (2019) Road verges support pollinators in agricultural landscapes, but are diminished by heavy traffic and summer cutting. *Journal of Applied Ecology* 56: 2316–2327.

Redon, L., Le Viol, I., Jiguet, F., Machon, N., Scher, O. & Kerbiriou, C. (2015) Road network in an agrarian landscape: potential habitat, corridor or barrier for small mammals? *Acta Oecologica* 62: 58–65.

Ruiz-Capillas, P., Mata, C. & Malo, J.E. (2013) Road verges are refuges for small mammal populations in extensively managed Mediterranean landscapes. *Biological Conservation* 158: 223–229.

Rytwinski, T. & Fahrig, L. (2007) Effect of road density on abundance of white-footed mice. *Landscape Ecology* 22: 1501–1512.

Rytwinski, T. & Fahrig, L. (2013) Why are some animal populations unaffected or positively affected by roads? *Oecologia* 173: 1143–1156.

Vermeulen, H. (1994) Corridor function of a road verge for dispersal of stenotopic heathland ground beetles Carabidae. *Biological Conservation* 69: 339–349.

Walthers, A.R. & Barber, C.A. (2020) Traffic noise as a potential stressor to offspring of an urban bird, the European starling. *Journal of Ornithology* 161: 459–467.

Whitworth, A., Beirne, C., Rowe, J., Ross, F., Acton, C., Burdekin, O. & Brown, P. (2015) The response of faunal biodiversity to an unmarked road in the Western Amazon. *Biodiversity and Conservation* 24: 1657–1670.

Wiącek, J., Kucharczyk, M., Polak, M. & Kucharczyk, H. (2014) Influence of road traffic on woodland birds – an experiment with using of nestboxes. *Sylwan* 158: 630–640.

Yosef, R. (2009) Highways as flyways: time and energy optimization in migratory Levant sparrowhawk. *Journal of Arid Environments* 73: 139–141.

Chapters 11 and 12

Al-Ghamdi, A.S. & AlGadhi, S.A. (2004) Warning signs as countermeasures to camel–vehicle collisions in Saudi Arabia. *Accident Analysis & Prevention* 36: 749–760.

Ament, R., Jacobson, S., Callahan, R. & Brocki, M. (eds.) (2021) *Highway Crossing Structures for Wildlife: Opportunities for Improving Driver and Animal Safety.* Albany, CA: US Department of Agriculture, Forest Service, Pacific Southwest Research Station.

Ascensão, F., Yogui, D.R., Alves, M.H., Alves, A.C., Abra, F. & Desbiez, A.L.J. (2021) Preventing wildlife roadkill can offset mitigation investments in short-medium term. *Biological Conservation* 253: 108902.

Benten, A., Annighöfer, P. & Vor, T. (2018) Wildlife warning reflectors' potential to mitigate wildlife-vehicle collisions—a review on the evaluation methods. *Frontiers in Ecology and Evolution* 6.

Berthinussen, A. & Altringham, J. (2012) Do bat gantries and underpasses help bats cross roads safely? *PLOS ONE* 7: e38775.

Bissonette, J.A., Kassar, C.A. & Cook, L.J. (2008) Assessment of costs associated with deer-vehicle collisions: human death and injury, vehicle damage, and deer loss. *Human–Wildlife Interactions*: 61.

Bissonette, J.A. & Silvia, R. (2012) An evaluation of a mitigation strategy for deer–vehicle collisions. *Wildlife Biology* 18: 414–423.

Bolliger, J., Hennet, T., Wermelinger, B., Bösch, R., Pazur, R., Blum, S., Haller, J. & Obrist, M.K. (2020) Effects of traffic-regulated street lighting on nocturnal insect abundance and bat activity. *Basic and Applied Ecology* 47: 44–56.

Buehler, R., Pucher, J., Gerike, R. & Götschi, T. (2017) Reducing car dependence in the heart of Europe: lessons from Germany, Austria, and Switzerland. *Transport Reviews* 37: 4–28.

Clevenger, A., Chruszcz, B. & Gunson, K. (2001) Highway mitigation fencing reduces wildlife–vehicle collisions. *Wildlife Society Bulletin* 29: 646–653.

Collinson, W.J., Marneweck, C. & Davies-Mostert, H.T. (2019) Protecting the protected: reducing wildlife roadkill in protected areas. *Animal Conservation* 22: 396–403.

Conan, A., Fleitz, J., Garnier, L., Le Brishoual, M., Handrich, Y. & Jumeau, J. (2022) Effectiveness of wire netting fences to prevent animal access to road infrastructures: an experimental study on small mammals and amphibians. *Nature Conservation* 47: 271–281.

Corlatti, L., Hackländer, K. & Frey-Roos, F. (2009) Ability of wildlife overpasses to provide connectivity and prevent genetic isolation. *Conservation Biology* 23: 548–56.

CPRE (2017) *The End of the Road? Challenging the Road-Building Consensus*. London: CPRE.

den Boer, L.C. & Schroten, A. (2007) *Traffic Noise Reduction in Europe: Health Effects, Social Costs and Technical and Policy Options to Reduce Road and Rail Traffic Noise*. Delft: CE Delft.

Denneboom, D., Bar-Massada, A. & Shwartz, A. (2021) Factors affecting usage of crossing structures by wildlife – a systematic review and meta-analysis. *Science of The Total Environment* 777: 146061.

Druta, C. & Alden, A.S. (2020) Preventing animal–vehicle crashes using a smart detection technology and warning system. *Transportation Research Record* 2674: 680–689.

Flatt, E., Basto, A., Pinto, C., Ortiz, J., Navarro, K., Reed, N., Brumberg, H., Chaverri, M.H. & Whitworth, A. (2022) Arboreal wildlife bridges in the tropical rainforest of Costa Rica's Osa Peninsula. *Folia Primatologica*: 1–17.

Found, R. & Boyce, M.S. (2011) Warning signs mitigate deer–vehicle collisions in an urban area. *Wildlife Society Bulletin* 35: 291–295.

Gaston, K.J., Davies, T.W., Bennie, J. & Hopkins, J. (2012) Reducing the ecological consequences of night-time light pollution: options and developments. *Journal of Applied Ecology* 49: 1256–1266.

Glazener, A., Wylie, J., van Waas, W. & Khreis, H. (2022) The impacts of car-free days and events on the environment and human health. *Current Environmental Health Reports* 9: 165–182.

Glista, D.J., DeVault, T.L. & DeWoody, J.A. (2009) A review of mitigation measures for reducing wildlife mortality on roadways. *Landscape and Urban Planning* 91: 1–7.

Gössling, S., Choi, A., Dekker, K. & Metzler, D. (2019) The social cost of automobility, cycling and walking in the European Union. *Ecological Economics* 158: 65–74.

Gunn, S. (2011) The Buchanan report, environment and the problem of traffic in 1960s Britain. *Twentieth Century British History* 22: 521–542.

Hall, P. (2004) The Buchanan report: 40 years on. *Proceedings of the Institution of Civil Engineers - Transport* 157: 7–14.

Handy, S. (2020) *Reducing Car Dependence Has Economic, Environmental, and Social Benefits*. Davis, CA: UC Davis, National Center for Sustainable Transportation.

Helldin, J.O. (2022) Are several small wildlife crossing structures better than a single large? Arguments from the perspective of large wildlife conservation. *Nature Conservation* 47: 197–213.

Hobday, A.J. (2010) Nighttime driver detection distances for Tasmanian fauna: informing speed limits to reduce roadkill. *Wildlife Research* 37: 265–272.

Huijser, M., Fairbank, E., Camel-Means, W., Graham, J., Watson, V., Basting, P. & Becker, D. (2016) Effectiveness of short sections of wildlife fencing and crossing structures along highways in reducing wildlife–vehicle collisions and providing safe crossing opportunities for large mammals. *Biological Conservation* 197: 61–68.

Huijser, M.P., Kociolek, A., McGowen, P., Hardy, A., Clevenger, A.P. & Ament, R. (2007) *Wildlife-Vehicle Collision and Crossing Mitigation Measures: A Toolbox for the Montana Department of Transportation*. Bozeman: Montana State University.

Innovate UK (2021) *UK TRANSPORT VISION 2050: Investing in the Future of Mobility*. Swindon: Innovate UK.

International Transport Forum (2021) *Reversing Car Dependency: Summary and Conclusions, ITF Roundtable Reports, No. 181*. Paris: OECD Publishing.

Jaarsma, C.F. & Willems, G.P.A. (2002) Reducing habitat fragmentation by minor rural roads through traffic calming. *Landscape and Urban Planning* 58: 125–135.

Jarvis, L.E., Hartup, M. & Petrovan, S.O. (2019) Road mitigation using tunnels and fences promotes site connectivity and population expansion for a protected amphibian. *European Journal of Wildlife Research* 65: 27.

Kociolek, A., Grilo, C. & Jacobson, S. (2015) Flight doesn't solve everything: mitigation of road impacts on birds. In: *Handbook of Road Ecology*, ed. R. Van der Ree, D.J. Smith & C. Grilo: pp. 281–289. John Wiley & Sons, Ltd.

Lesbarrères, D. & Fahrig, L. (2012) Measures to reduce population fragmentation by roads: what has worked and how do we know? *Trends in Ecology & Evolution* 27: 374–380.

Lister, N.-M., Brocki, M. & Ament, R. (2015) Integrated adaptive design for wildlife movement under climate change. *Frontiers in Ecology and the Environment* 13: 493–502.

Mata, C., Hervás, I., Herranz, J., Suárez, F. & Malo, J.E. (2008) Are motorway wildlife passages worth building? Vertebrate use of road-crossing structures on a Spanish motorway. *Journal of Environmental Management* 88: 407–415.

Matos, C., Petrovan, S., Ward, A.I. & Wheeler, P. (2017) Facilitating permeability of landscapes impacted by roads for protected amphibians: patterns of movement for the great crested newt. *PeerJ* 5: e2922.

Mattioli, G., Roberts, C., Steinberger, J.K. & Brown, A. (2020) The political economy of car dependence: a systems of provision approach. *Energy Research & Social Science* 66: 101486.

McCollister, M.F. & Van Manen, F.T. (2010) Effectiveness of wildlife underpasses and fencing to reduce wildlife-vehicle collisions. *The Journal of Wildlife Management* 74: 1722–1731.

Mitchell, P. (2009) *Speed and Road Traffic Noise: A Report to the UK Noise Association*. Chatham: UK Noise Association.

Moriarty, P. (2022) Electric vehicles can have only a minor role in reducing transport's energy and environmental challenges. *AIMS Energy* 10: 131–148.

Natural England Commissioned Report NECR181 (2015) *Green Bridges: A Literature Review*. UK: Natural England.

Niehaus, A.C. & Wilson, R.S. (2018) Integrating conservation biology into the development of automated vehicle technology to reduce animal–vehicle collisions. *Conservation Letters* 11: e12427.

Okita-Ouma, B., Koskei, M., Tiller, L., Lala, F., King, L., Moller, R., Amin, R. & Douglas-Hamilton, I. (2021) Effectiveness of wildlife underpasses and culverts in connecting elephant habitats: a case study of new railway through Kenya's Tsavo National Parks. *African Journal of Ecology* 59: 624–640.

Oliveira Gonçalves, L., Kindel, A., Augusto Galvão Bastazini, V. & Zimmermann Teixeira, F. (2022) Mainstreaming ecological connectivity in road environmental impact assessments: a long way to go. *Impact Assessment and Project Appraisal*: 1–6.

Ottburg, F.G.W.A. & van der Grift, E.A. (2019) Effectiveness of road mitigation for common toads (*Bufo bufo*) in the Netherlands. *Frontiers in Ecology and Evolution,* 7, 10.3389/fevo.2019.00023.

Phillips, B.B., Wallace, C., Roberts, B.R., Whitehouse, A.T., Gaston, K.J., Bullock, J.M., Dicks, L.V. & Osborne, J.L. (2020) Enhancing road verges to aid pollinator conservation: a review. *Biological Conservation* 250: 108687.

Plaschke, M., Bhardwaj, M., König, H., Wenz, E., Dobiáš, K. & Ford, A. (2021) Green bridges in a re-colonizing landscape: wolves (*Canis lupus*) in Brandenburg, Germany. *Conservation Science and Practice* 3: DOI:10.1111/csp2.364.

Polak, T., Nicholson, E., Grilo, C., Bennett, J. & Possingham, H. (2018) Optimal planning to mitigate the impacts of roads on multiple species. *Journal of Applied Ecology* 56: 201–213.

Rhodes, J.R., Lunney, D., Callaghan, J. & McAlpine, C.A. (2014) A few large roads or many small ones? How to accommodate growth in vehicle numbers to minimise impacts on wildlife. *PLOS ONE* 9: e91093.

Rice, W.L., Newman, P., Zipp, K.Y., Taff, B.D., Pipkin, A.R., Miller, Z.D. & Pan, B. (2022) Balancing quietness and freedom: support for reducing road noise among park visitors. *Journal of Outdoor Recreation and Tourism* 37: 100474.

Riginos, C., Fairbank, E., Hansen, E., Kolek, J. & Huijser, M.P. (2022) Reduced speed limit is ineffective for mitigating the effects of roads on ungulates. *Conservation Science and Practice* 4: e618.

Rytwinski, T., Soanes, K., Jaeger, J.A.G., Fahrig, L., Findlay, C.S., Houlahan, J., van der Ree, R. & van der Grift, E.A. (2016) How effective is road mitigation at reducing road-kill? A meta-analysis. *PLOS ONE* 11: e0166941.

Sawaya, M.A., Kalinowski, S.T. & Clevenger, A.P. (2014) Genetic connectivity for two bear species at wildlife crossing structures in Banff National Park. *Proceedings of the Royal Society B: Biological Sciences* 281: 20131705.

Silva, Ó., Cordera, R., González-González, E. & Nogués, S. (2022) Environmental impacts of autonomous vehicles: a review of the scientific literature. *Science of the Total Environment*: 154615.

Sloman, L., Hopkinson, L. & Taylor, I. (2017) *The Impact of Road Projects in England. Report for CPRE*. London: Transport for Quality of Life.

Smith, D.J., van der Ree, R. & Rosell, C. (2015) Wildlife crossing structures: An effective strategy to restore or maintain wildlife connectivity across roads. In: *Handbook of Road Ecology,* ed. R. van der Ree, D.J. Smith & C. Grilo: pp. 172–183. UK: John Wiley & Sons.

Soanes, K., Taylor, A.C., Sunnucks, P., Vesk, P.A., Cesarini, S. & van der Ree, R. (2018) Evaluating the success of wildlife crossing structures using genetic approaches and an experimental design: lessons from a gliding mammal. *Journal of Applied Ecology* 55: 129–138.

Spanowicz, A.G., Teixeira, F.Z. & Jaeger, J.A.G. (2020) An adaptive plan for prioritizing road sections for fencing to reduce animal mortality. *Conservation Biology* 34: 1210–1220.

Sullivan, T.L., Williams, A.F., Messmer, T.A., Hellinga, L.A. & Kyrychenko, S.Y. (2004) Effectiveness of temporary warning signs in reducing deer–vehicle collisions during mule deer migrations. *Wildlife Society Bulletin* 32: 907–915.

Switalski, T.A., Bissonette, J.A., DeLuca, T.H., Luce, C.H. & Madej, M.A. (2004) Benefits and impacts of road removal. *Frontiers in Ecology and the Environment* 2: 21–28.

Tryjanowski, P., Beim, M., Kubicka, A.M., Morelli, F., Sparks, T.H. & Sklenicka, P. (2021) On the origin of species on road warning signs: a global perspective. *Global Ecology and Conservation* 27: e01600.

van der Grift, E.A., van der Ree, R., Fahrig, L., Findlay, S., Houlahan, J., Jaeger, J.A.G., Klar, N., Madrinan, L.F. & Olson, L. (2013) Evaluating the effectiveness of road mitigation measures. *Biodiversity and Conservation* 22: 425–448.

Van Renterghem, T. (2014) Guidelines for optimizing road traffic noise shielding by non-deep tree belts. *Ecological Engineering* 69: 276–286.

Zhang, W., Shu, G., Li, Y., Xiong, S., Liang, C. & Li, C. (2018) Daytime driving decreases amphibian roadkill. *PeerJ* 6: e5385.

COVID-19 and the anthropause

Aletta, F., Oberman, T., Mitchell, A., Tong, H. & Kang, J. (2020) Assessing the changing urban sound environment during the COVID-19 lockdown period using short-term acoustic measurements. *Noise Mapping* 7: 123–134.

Basak, S.M., O'Mahony, D.T., Lesiak, M., Basak, A.K., Ziółkowska, E., Kaim, D., Hossain, M.S. & Wierzbowska, I.A. (2022) Animal–vehicle collisions during the COVID-19 lockdown in early 2020 in the Krakow metropolitan region, Poland. *Scientific Reports* 12: 7572.

Bekbulat, B., Apte, J.S., Millet, D.B., Robinson, A.L., Wells, K.C., Presto, A.A. & Marshall, J.D. (2021) Changes in criteria air pollution levels in the US before, during, and after Covid-19 stay-at-home orders: evidence from regulatory monitors. *Science of the Total Environment* 769: 144693.

Bíl, M., Andrášik, R., Cícha, V., Arnon, A., Kruuse, M., Langbein, J., Náhlik, A., Niemi, M., Pokorny, B., Colino-Rabanal, V.J., Rolandsen, C.M. & Seiler, A. (2021) COVID-19 related travel restrictions prevented numerous wildlife deaths on roads: a comparative analysis of results from 11 countries. *Biological Conservation* 256: 109076.

Chang, H.-H., Meyerhoefer, C.D. & Yang, F.-A. (2021) COVID-19 prevention, air pollution and transportation patterns in the absence of a lockdown. *Journal of Environmental Management* 298: 113522.

de-Albéniz, Í.G.-M., de-Villa, J.A.R. & Rodriguez-Hernandez, J. (2022) Impact of COVID-19 lockdown on wildlife–vehicle collisions in NW of Spain. *Sustainability* 14: 1–14.

Elshorbany, Y.F., Kapper, H.C., Ziemke, J.R. & Parr, S.A. (2021) The status of air quality in the United States during the COVID-19 pandemic: a remote sensing perspective. *Remote Sensing* 13: 369.

Faridi, S., Yousefian, F., Janjani, H., Niazi, S., Azimi, F., Naddafi, K. & Hassanvand, M.S. (2021) The effect of COVID-19 pandemic on human mobility and ambient air quality around the world: a systematic review. *Urban Climate* 38: 100888.

Garg, N. (2021) Impact of COVID-19 lockdown on ambient noise levels in seven metropolitan cities of India. *Applied Acoustics*: 108582.

Hicks, W., Beevers, S., Tremper, A., Stewart, G., Priestman, M., Kelly, F., Lanoisellé, M., Lowry, D. & Green, D. (2021) Quantification of non-exhaust particulate matter traffic emissions and the impact of COVID-19 lockdown at London Marylebone Road. *Atmosphere* 12: 1–19.

Hudda, N., Simon, M.C., Patton, A.P. & Durant, J.L. (2020) Reductions in traffic-related black carbon and ultrafine particle number concentrations in an urban neighborhood during the COVID-19 pandemic. *Science of the Total Environment* 742: 140931.

Jephcote, C., Hansell, A.L., Adams, K. & Gulliver, J. (2021) Changes in air quality during COVID-19 'lockdown' in the United Kingdom. *Environmental Pollution* 272: 116011.

Kerr, G.H., Goldberg, D.L. & Anenberg, S.C. (2021) COVID-19 pandemic reveals persistent disparities in nitrogen dioxide pollution. *Proceedings of the National Academy of Sciences* 118: e2022409118.

LeClair, G., Chatfield, M.W.H., Wood, Z., Parmelee, J. & Frederick, C.A. (2021) Influence of the COVID-19 pandemic on amphibian road mortality. *Conservation Science and Practice* 3: e535.

Loh, H.C., Looi, I., Ch'ng, A.S.H., Goh, K.W., Ming, L.C. & Ang, K.H. (2021) Positive global environmental impacts of the COVID-19 pandemic lockdown: a review. *GeoJournal.*

Łopucki, R., Kitowski, I. & Klich, D. (2021) How is wildlife affected by the COVID-19 pandemic? Lockdown effect on the road mortality of hedgehogs. *Animals* 11: 868.

Maji, K.J., Namdeo, A., Bell, M., Goodman, P., Nagendra, S.M.S., Barnes, J.H., De Vito, L., Hayes, E., Longhurst, J.W., Kumar, R., Sharma, N., Kuppili, S.K. & Alshetty, D. (2021) Unprecedented reduction in air pollution and corresponding short-term premature mortality associated with COVID-19 lockdown in Delhi, India. *Journal of the Air & Waste Management Association* 71: 1085–1101.

Manenti, R., Mori, E., Di Canio, V., Mercurio, S., Picone, M., Caffi, M., Brambilla, M., Ficetola, G.F. & Rubolini, D. (2020) The good, the bad and the ugly of COVID-19 lockdown effects on wildlife conservation: insights from the first European locked down country. *Biological Conservation* 249: 108728.

Mimani, A. & Singh, R. (2021) Anthropogenic noise variation in Indian cities due to the COVID-19 lockdown during March-to-May 2020. *The Journal of the Acoustical Society of America* 150: 3216–3227.

Nieuwenhuijsen, M.J., Hahad, O. & Münzel, T. (2021) The COVID-19 pandemic as a starting point to accelerate improvements in health in our cities through better urban and transport planning. *Environmental Science and Pollution Research.*

Pokorny, B., Cerri, J. & Bužan, E. (2022) Wildlife roadkill and COVID-19: a biologically significant, but heterogeneous, reduction. *Journal of Applied Ecology*, https://doi.org/10.1111/1365-2664.14140.

Schrimpf, M.B., Brisay, P.G.D., Johnston, A., Smith, A.C., Sánchez-Jasso, J., Robinson, B.G., Warrington, M.H., Mahony, N.A., Horn, A.G., Strimas-Mackey, M., Fahrig, L. & Koper, N. (2021) Reduced human activity during COVID-19 alters avian land use across North America. *Science Advances* 7: DOI: 10.1126/sciadv.abf5073

Shilling, F. (2020) *Special Report 3: Impact of COVID19 Mitigation on Traffic, Fuel Use and Climate Change.* Davis, CA: Road Ecology Centre, UC Davis.

Shilling, F., Nguyen, T., Saleh, M., Kyaw, M.K., Tapia, K., Trujillo, G., Bejarano, M., Waetjen, D., Peterson, J., Kalisz, G., Sejour, R., Croston, S. & Ham, E. (2021) A reprieve from US wildlife mortality on roads during the COVID-19 pandemic. *Biological Conservation* 256: 109013.

Venter, Z.S., Aunan, K., Chowdhury, S. & Lelieveld, J. (2020) COVID-19 lockdowns cause global air pollution declines. *Proceedings of the National Academy of Sciences* 117: 18984.

Weir, B., Crisp, D., O'Dell, C.W., Basu, S., Chatterjee, A., Kolassa, J., Oda, T., Pawson, S., Poulter, B., Zhang, Z., Ciais, P., Davis, S.J., Liu, Z. & Ott, L.E. (2021) Regional impacts of COVID-19 on carbon dioxide detected worldwide from space. *Science Advances* 7: eabf9415.

Index

Also available from Pelagic

Reflections: What Wildlife Needs and How to Provide It, Mark Avery (coming autumn 2023)

Reconnection: Fixing Our Broken Relationship with Nature, Miles Richardson

Treated Like Animals: Improving the Lives of the Creatures We Own, Eat and Use, Alick Simmons

Invisible Friends: How Microbes Shape Our Lives and the World Around Us, Jake M. Robinson

Low-Carbon Birding, edited by Javier Caletrío

The Hen Harrier's Year, Ian Carter and Dan Powell

Wildlife Photography Fieldcraft: How to Find and Photograph UK Wildlife, Susan Young

Rhythms of Nature: Wildlife and Wild Places Between the Moors, Ian Carter

Ancient Woods, Trees and Forests: Ecology, Conservation and Management, edited by Alper H. Çolak, Simay Kırca and Ian D. Rotherham

Essex Rock: Geology Beneath the Landscape, Ian Mercer and Ros Mercer

Pollinators and Pollination, Jeff Ollerton

Wild Mull: A Natural History of the Island and its People, Stephen Littlewood and Martin Jones

Challenges in Estuarine and Coastal Science, edited by John Humphreys and Sally Little

A Natural History of Insects in 100 Limericks, Richard A. Jones and Calvin Ure-Jones